WHY GEOGRAPHY MATTERS

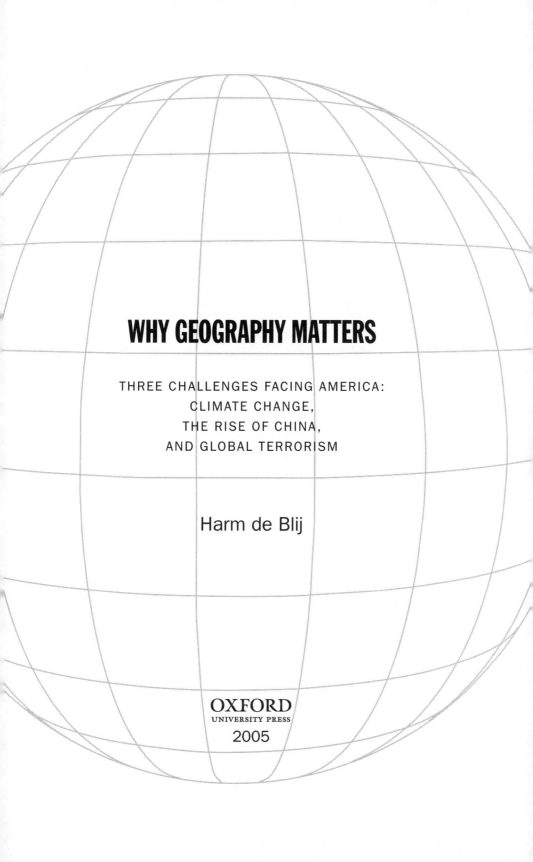

WHY GEOGRAPHY MATTERS

THREE CHALLENGES FACING AMERICA:
CLIMATE CHANGE,
THE RISE OF CHINA,
AND GLOBAL TERRORISM

Harm de Blij

OXFORD
UNIVERSITY PRESS
2005

OXFORD
UNIVERSITY PRESS

Oxford University Press, Inc., publishes works that further
Oxford University's objective of excellence
in research, scholarship, and education.

Oxford New York
Auckland Cape Town Dar es Salaam Hong Kong Karachi
Kuala Lumpur Madrid Melbourne Mexico City Nairobi
New Delhi Shanghai Taipei Toronto

With offices in
Argentina Austria Brazil Chile Czech Republic France Greece
Guatemala Hungary Italy Japan Poland Portugal Singapore
South Korea Switzerland Thailand Turkey Ukraine Vietnam

Published by Oxford University Press, Inc.
198 Madison Avenue, New York, New York, 10016
www.oup.com

Library of Congress Cataloging-in-Publication Data
De Blij, Harm J.
Why geography matters : three challenges facing America : Climate change, the rise of China,
and global terrorism / H. J. de Blij.
p. cm.
ISBN-13: 978–0–19–518301–6
ISBN-10: 0–19–518301–0
1. Human geography—United States—History—21st century.
2. Terrorism—History—21st century.
3. Climate change—History—21st century.
4. China—Politics and government—2002–
5. United States—Politics and government—2001–
6. United States—Social conditions—21st century.
Title.
GF503.D4 2005
909.83—dc22   2004030369

Maps drawn by Mapping Specialists, Madison, Wisconsin

3 5 7 9 8 6 4 2
Printed in the United States of America
on acid-free paper

To Macduff, Cleo, and Barley,
who are sure to chew on this.

CONTENTS

PREFACE

Ten years ago I wrote a book about the state of the world from a geographic perspective, guided in part by the preferences of viewers who had watched my appearances on ABC's *Good Morning America* over the preceding seven years (de Blij, 1995). We had mapped on screen the collapse of the Soviet Union, Iraq's invasion of Kuwait, the first Gulf War, the breakup of Yugoslavia, the economic miracle on the Pacific Rim, the apparent strides the planet was making toward a New World Order. There was still hope that global warming would be mild and temporary, that Africa's health and economic malaise would reverse, that Russia was headed for stable democracy, that incidents of terrorism would decline in number and severity.

What a difference a decade makes in this fast-changing world. Abrupt climate change has become a real and serious threat to global order. China's economic rise is shadowed by political clout and military might that has the potential to precipitate a new Cold War. Terrorism has intensified and is taking on the character of a global insurgency. Europe's unification is faltering; Russia's democratization is failing. And Africa's prospects, in this era of globalization, have deteriorated, not improved.

Not since the days when the sun never set over the British Empire has one superpower had the global influence deployed today by the United States of America. When foreign observers stated in their dispatches that the November 2004 United States presidential election was in effect an election for world leadership as well, they did not exaggerate. The rest of the world did not have a vote, but it had reasons to take an interest in the result. At the turn of the century, United States power was an acknowledged reality, but repeated assaults on American targets at home and abroad had been met with tepid responses. United States relations with foreign countries still were generally good. Four years later, America's forces had occupied two Muslim countries and the American government left no doubt that others, Muslim and non-Muslim, were in the crosshairs. American dominance was, by some measures, at an all-time high. But United States relations with many foreign countries were at a historic low.

As the democratic nation that elects representatives whose decisions affect not just America but the entire world, we Americans have an obligation to be well informed about our small and functionally shrinking planet. But the evidence is that our geographic literacy is inadequate and that our interest is lagging—at least when it comes to world affairs. Between the early 1980s and the start of the war in Iraq, international news coverage, measured in minutes devoted to it annually, fell by two-thirds.

That is the point of this book: to argue that geographic literacy is a matter of national security, so that the weak state of our geographic knowledge constitutes a serious, perhaps critical, disadvantage in an increasingly competitive world. Geographic insights can be crucial in addressing geopolitical problems; they are needed also in decision making in spheres ranging from the cultural to the economic. We need to make better use of maps, to exploit the new technologies that have transformed cartography from illustration to interaction. We need to be more familiar with the globe and what, after all these years, it can still teach us about direction and orientation.

One of my colleagues, to whom I had sent several draft chapters for review and opinion, asked me what my hopes are for this book. Most of all I hope that it will confirm the vitality and utility of geography as a way to understand our complex world. Geography tends to come up with unexpected linkages—between climate change and historical events, between natural phenomena and political developments, between environment and behavior—that are unmatched in other fields. And geography tends also to look at the here and now, and perceptively into the future. Coupled with Americans' low level of geographic literacy is a kind of cultural obsession with the past. It is inconceivable to go through college without courses in history, archeology, geology, and others of chronological bent. But few of us graduate with some exposure to modern geography—far too few.

The other hope I have is that teachers, whether geography teachers or others, will find in this book topics to discuss with their students, and that parents will find issues here to discuss with their families around the dinner table. If I may make a suggestion: when reading this book, please have an atlas handy. This book contains maps, but you can never have too many. And a globe would be useful as well. No home should be without one. Should you find yourself wanting to delve more deeply into geography, a textbook listed in the citations contains more than 200 thematic color maps drawn by Mapping Specialists, whose excellent work has supported my writings for many years (de Blij and Muller, 2004).

I had the good fortune to encounter geography as a future profession early in life, and that good fortune was compounded by the colleagues I met and the friends I made among the small but vibrant community of geographers working in a range of professional settings that would astound you (from the wine industry to NOAA, from the travel industry to teaching, from the publishing industry to government). Many of these colleagues and friends, some in related disciplines, contributed in one way or another to this book, by bringing developments and ideas to my attention, by debating content, by reading drafts, sometimes by making a brief comment that led to unanticipated insights. Some of this assistance is reflected by the citations, but here I express my appreciation to Dick Allen, Barbara Bailey, Richard Banz, Bob Begam, Phil Benton, Alan Best, Dick Boehm, Dave Campbell, Iraphne Childs, Spencer Christian, Roy Cole, Bob Crook, Wade Currier, George Demko, John Drinko, Ryan Flahive, Skip Fleischmann, Erin Fouberg, Gary Fuller, Charlie Gibson, Ed Grode, Dick Groop, Gil Grosvenor, Jay Harman, Carl Hirsch, Chip Hoehler, John Hunter, Tom Jeffs, Arte Johnson, Ted Kandle, Grady Kelly-Post, Jim King, Peter Krogh, John Larsen, Neal Lineback, Don Lubbers, John Malinowski, Geof Martin, Patricia McCulley, Dave Mc-Farland, Paul McFarland, Betty Meggers, Hugh Moulton, Peter Muller, Alec Murphy, David Murray, Jim Newman, Dan Niemeyer, Jan Nijman, Eugene Palka, Charlie Pirtle, Mac Pohn, Jim Potchen, Frank Potter, Victor Prescott, Doug Richardson, Jack Reilly, Dick Robb, Chris Rogers, Eunice Rutledge, Stathis Sakellarios, Birthe Sauer, Jeff Schaffer, Ruth Sivard, Thomas Sowell, Peter Spiers, Marc Swartz, Rich Symanski, Mary Swingle, Derrick Thom, Morris Thomas, George Valanos, Jack Weatherford, Frank Whitmore, Antoinette WinklerPrins, Duke Winters, and Henry Wright. I am, of course, solely responsible for the contents of this book, and nothing in it implies endorsement by all or any of those just cited.

I am especially grateful to my editor at Oxford University Press, Ben Keene, who proposed that I write this book and whose intense interest in it was matched by his creative energies; every author should be so fortunate. I further acknowledge appreciatively the superb work of Catherine Humphries, the production editor, the constructive interventions of the copyeditor, Paula Cooper, and the creative layouts of the designer, Susan Day, and Kathleen Lynch, who created the attractive cover. I also thank my publicist, Jordan Bucher, and acknowledge the generosity of William B. Kory, editor of the *Pennsylvania Geographer*, who allowed me to modify text from my article on Africa (de Blij, 2004) for inclusion in chapter 12.

It is always difficult to present complex information on black-and-white maps, and I am grateful to Don Larson and his team at Mapping Specialists in Madison, Wisconsin for their splendid work as they made the most of limited cartographic options.

Most of all I thank my wife, Bonnie, for her encouragement, unfailing good judgment, and forbearance. No brief acknowledgment of her role in making this effort possible could begin to reflect the measure of my gratitude.

Harm de Blij

WHY GEOGRAPHY MATTERS

1

WHY GEOGRAPHY MATTERS

Ten years ago it seemed that the world could not possibly change any faster than it had over the previous decade. The Soviet Union had disintegrated into 15 newly independent countries, China's Pacific Rim was transforming the economic geography of East Asia, South Africa was embarked on a new course under the guidance of Nelson Mandela (a new course that relied on the complete reconstruction of its administrative map), NAFTA linked Canada, the United States, and Mexico in an economic union that would change the commercial map of North America, the European Community was renamed the European Union and added three members to its roster to create a 15-nation entity, and Yugoslavia was collapsing amid uncontrolled carnage. All this change in the human geography of the world was matched by permutations of nature ranging from global climate change to local environmental extremes among which the "thousand-year" flooding of the Mississippi-Missouri river systems of 1993 was but one of the many dramatic events. New terms came into general use: ethnic cleansing, greenhouse warming, Gulf War, El Niño.

Yet the pace of change during the decade straddling the turn of the century has not slowed down. The widening circle of international terrorism has invaded the United States in once-unimaginable ways. American troops fought in Afghanistan and invaded Iraq. North Korea and its nuclear ambitions rose to the forefront of international concerns. Almost-overlooked wars in Africa cost millions—yes, millions—of lives, and the growing spread of AIDS cost millions more. Jobs lost by the United States and Canada to Mexico through opportunities created by NAFTA began to siphon off to China. The European Union continued to expand, taking in ten new members in May 2004 and incorporating 25 of Europe's 39 countries with others waiting in the wings. The name Yugoslavia disappeared from the map even as others emerged: East Timor, Papua, Padania, Transdniestria, Limpopo, Uttaranchal. And new terms in common usage reflected the new era: pandemic, *jihad*, War on Terrorism, Sunni Triangle.

Is there a common denominator for all this change? Can our world and its transformations be better appreciated through a particular perspective? This book answers both questions with one affirmation: geography.

In truth, geography itself has gone through several transformations in recent times. When I was a high-school student, learning to name countries and cities, ranges and rivers was an end in itself. Making the connections that give geography its special place among the sciences was not on the agenda. By the time I got to college, geography (in Europe and America at least) had become more scientific, even mathematical. During my teaching career it became more technological, and not for nothing does the now-common acronym GIS stand for Geographic Information Systems. Today geography has numerous dimensions, but it remains a great way to comprehend our complex world.

BECOMING A GEOGRAPHER

The other day I read an interview with a prominent geographer in the newsletter of this country's largest professional geographic organization. The editor asked Frederick E. (Fritz) Nelson, now teaching at the University of Delaware, a question all of us geographers hear often: what caused you to join our ranks? His answer is one given by many a colleague: while an undergraduate (at Northern Michigan University) he took a course in regional geography and liked it so much that he decided to pursue a major in the discipline. He changed directions while a graduate student at Michigan State University, but he did not forget what attracted him to geography originally. Today his research on the geography of periglacial (ice-margin) phenomena is world renowned (Solis, 2004).

My own encounter with geography stems from my very first experience with it in Holland during the Second World War, not at school, but at home. With my dad I watched in horror from a roof window in our suburban house when my city, Rotterdam, was engulfed by flames following the Nazi firebombing of May 14, 1940 (long-buried feelings that resurfaced on September 11, 2001), and soon my parents abandoned the city for a small village near the center of the country. There they engaged in a daily battle for survival, and I spent much time in their library, which included several world and national atlases, a large globe, and the books of a geographer named Hendrik Willem van Loon. As the winters grew colder and our situation deteriorated, those books gave me hope. Van Loon described worlds far away, where it was warm, where skies were blue and palm trees swayed in soft breezes, and where food could be plucked from trees. There were exciting descriptions of active volcanoes and of tropical storms, of maritime journeys to remote islands, of great, bustling cities, of powerful kingdoms and unfamiliar customs. I traced van Loon's journeys on atlas maps and dreamed of the day when I would see his worlds for myself. Van Loon's geographies gave me, almost literally, a lease on life.

After the end of the war, my fortunes changed in more than one way. When the schools opened again, my geography teacher was an inspiring taskmaster who made sure that we, a classroom full of youngsters with a wartime gap in our early education, learned that while geography could widen our horizons, it also required some rigorous studying. The rewards, he rightly predicted, were immeasurable.

If, therefore, I write of geography with enthusiasm and in the belief that it can make life easier and more meaningful in this complex and changing world, it is because of a lifetime of discovery and fascination.

WHAT IS GEOGRAPHY?

As a geographer, I've often envied my colleagues in such fields as history, geology, and biology. It must be wonderful to work in a discipline so well defined by its name and so accurately perceived by the general public. Actually, the public's perception may not be so accurate, but people *think* they know what historians, geologists, and biologists do.

We geographers are used to it. Sit down next to someone in an airplane or in a waiting room somewhere, get involved in a conversation, and that someone is bound to ask: Geography? You're a geographer? What is geography, anyway?

In truth, we geographers don't have a single, snappy answer. A couple of millennia ago, geography essentially was about discovery. A Greek philosopher named Eratosthenes moved geographic knowledge forward by leaps and bounds; by measuring Sun angles, he not only concluded that the Earth was round but came amazingly close to the correct figure for its circumference. Several centuries later, geography was propelled by exploration and cartography, a period that came to a close, more or less, with the adventures and monumental writings of Alexander von Humboldt, the German naturalist-geographer. A few decades ago, geography still was an organizing, descriptive discipline whose students were expected to know far more capes and bays than were really necessary. Today geography is in a new technological age, with satellites transmitting information to computers whose maps are used for analysis and decision making.

Despite these new developments, however, geography does have some traditions. The first, and in many ways the most important, is that geography deals with the natural as well as the human world. It is, therefore, not just a "social" science. Geographers do research on glaciations and coastlines, on desert dunes and limestone caves, on weather and climate, even on plants and animals. We also study human activities, from city planning to boundary making, from wine growing to churchgoing. To me, that's the best part of geography: there's almost nothing in this wide, wonderful world of ours that can't be studied geographically.

This means, of course, that geographers are especially well placed to assess the complicated relationships between human societies and natural environments; this is geography's second tradition. In this arena knowledge is fast growing, and if you want to see evidence of the insights geography can contribute I know of no better book than Jean Grove's spellbinding analysis of what happened when Europe and much of the rest of the world were plunged into what she calls *The Little Ice Age*, starting around 1300 and continuing, with a few letups, until the early 1800s. This is a global, sweeping analysis; other geographers work at different levels of scale. Some of my colleagues study and predict people's reactions to environmental hazards: Why do people persist in living on the slopes of active volcanoes and in the floodplains of flood-prone rivers? How much do home buyers in California know about the earthquake risk at the location of their purchase and what are they told by real-estate agents before they buy? Another environment-related issue involves health and disease. The origins and spread of many diseases have much to do with climate, vegetation, and fauna as well as cultural traditions and habits. A small but productive cadre of medical geographers is at work researching and predicting outbreaks and dispersals of maladies ranging from cholera to AIDS to bird flu. Peter Gould's book on AIDS, which he called *The Slow Plague*, effectively displays the toolbox of geographers when it comes to such analyses (Gould, 1993).

A third geographic tradition is simply this: we do research in, and try to understand, foreign cultures and distant regions. A few decades ago, it was rare to find a geographer who did not have some considerable expertise in a foreign area, large or small. Most spoke one or more foreign languages (this used to be a requirement for graduation with a doctorate), kept up with the scholarly literature as well as the popular press in their chosen region, and conducted repeated research there. That tradition has faded somewhat in the new age of the Internet, satellite data, and computer cartography, but many students still are first attracted to geography because it aroused their curiosity about some foreign place. The decline in interest in international affairs is not unique to geography, of course. From analyses of network news content to studies of foreign-area specialization in United States intelligence operations, our isolationism and parochialism are evident. But there will be a rebound, probably of necessity more than desire. Geographic provincialism entails serious national security risks.

A fourth tradition geographers like to identify is the so-called location tradition, which is essentially a human-geographic (not a physical-geographic) convention. Why are activities, such as movie industries or shopping centers, or towns or cities such as Sarasota, Florida, or Tokyo, Japan located where they are? What does their location imply for their prospects? Why did one city thrive and grow while a nearby settlement dwindled and failed? Often a geographic answer illuminates historic events. Urban and regional planning is now a key component to many college geography curricula, and many of our graduates find positions in the planning field.

Having said all this, we still don't have a short-answer definition of geography.

LOOKING AT THE WORLD SPATIALLY

To pull it all together, we need a word that telegraphs our main geographic preoccupation, and that word derives from space—not celestial space, but Earthly space. We geographers look at the world *spatially*. I sometimes try this concept on a questioner: historians look at the world temporally or chronologically; economists and political scientists come at it structurally, but we geographers look at it spatially. With a little luck my interrogator will furrow his brow, nod understandingly, and take out his *USA Today* and read about the results of the latest geographic literacy test.

Geographers, of course, are not the only scholars to use spatial analysis to explain the workings of our world. Economists, anthropologists, and other social scientists sometimes take a spatial perspective as well although, as their writings suggest, they often lag behind. Geographers were amused (a few were annoyed) when the noted economist Paul Krugman began writing his columns in the *New York Times* and rediscovered spatial truisms that had long since been superseded in the geographic literature. The physiologist Jared Diamond's magisterial book *Guns, Germs, and Steel* was described by *New York Times* journalist John N. Wilford as "the best book on geography in recent years," but geographers noted some significant conceptual weaknesses in it (Diamond, 1997). Mr. Diamond not only took note of these caveats, but acted impressively on them: he joined the faculty of the Department of Geography at UCLA and wrote a successor volume that demonstrates his perception of geographic factors in the disintegration of once-thriving societies (Diamond, 2005).

Diamond, in both of these Herculean works, raises sensitive issues that once lay at the core of geographic research: the role of natural environments in the fate of human societies. Early in the twentieth century, this research led to generalizations attributing the "energy" of midlatitude societies and the "lethargy" of tropical peoples to climate. Such simplistic analyses were not only bound to be flawed, but could be used to give credence to racist interpretations of the state of the world, discrediting the whole enterprise. But the fundamental question, as Diamond asserts, has not gone away. Today we know a great deal more about environmental swings and associated ecological transitions as well as human dispersal and behavior, and the issue is getting renewed attention.

Nevertheless, it remains tempting to assign a simple causal relationship to a complex set of circumstances because a map suggests it. Consider the following quote from a lecture presented at the United States Naval War College by another noted economist, Jeffrey Sachs: "Virtually all of the rich countries of the world are outside the tropics, and virtually all of the poor countries are within them . . . climate, then, accounts for quite a significant proportion of the cross-national and cross-regional disparities of world income" (Sachs, 2000). That would seem to be a reasonable conclusion, but the condition of many of the world's poor countries results from a far more complex combination of circumstances including subjugation, colonialism, exploitation, and suppression that put them at a disadvantage that will long endure and for which climate may not be the significant causative factor Mr. Sachs implies. In any case, while it is true that many of the world's poor countries lie in tropical environs, many others, from Albania to Turkmenistan and from Moldova to North Korea, do not. The geographic message does not lend itself to environmental generalizations.

Of course we should be pleased that nongeographers are jumping on our bandwagon, but this does not make our effort to come up with a generally accepted definition of our discipline any easier. In some ways, I suppose, this very difficulty is one of geography's strengths. Geography is a discipline of diversity, under whose "spatial" umbrella we study and analyze processes, systems, behaviors, and countless other phenomena that have spatial expression. It is this tie that binds geographers, this interest in patterns, distributions, diffusions, circulations, interactions, juxtapositions—the ways in which the physical and human worlds are laid out, interconnect, and interact. Yes, it is true that some tropical environs are tough on farmers and engender diseases. Tougher still, though, are the rich world's tariff barriers against the produce of tropical-country farmers and the subsidies paid to large agribusiness. End those practices, and suddenly climate won't seem so "significant" a factor in the global distribution of poverty.

So geography's umbrella is large, allowing geographers to pursue widely varying research. These days that includes a lot of social activism and other work that might seem closer to sociology than to geography, but much geographic research remains spatial and substantive. I have colleagues whose work focuses on Amazonian deforestation, West African desertification, Asian economic integration, Indonesian transmigration. Others take a more specific look at such American phenomena as professional football and the sources and team destinations of players, the changing patterns of church membership and evangelism, the rise of the wine industry in this period of global warming, and the impact of NAFTA on manufacturing employment in the Midwest. I'm always fascinated to read in our professional journals what they're discovering, and as I used to tell my students, the Age of Discovery may be over, but the era of geographic discovery never will be.

THE SPATIAL SPECIALIZERS

The stirring story of geography's early emergence, its Greek and Roman expansion, its European diversification, and its global dissemination is a saga of pioneering observation, heroic exploration, inventive mapmaking, and ever-improving interpretation, discussed in fascinating detail by the discipline's leading historian (Martin, 2005). Long before European colonialism launched the first wave of what today we call globalization, indigenous geographers were drawing maps and interpreting landscapes from Korea to the Andes and from India to Morocco. Later, geographic philosophy got caught up in European nationalism, and various "schools" of geography—German, French, British—came to reflect, and even to support and justify, national political and strategic aspirations including expansionism, colonialism, and even Nazism. In the United States, geography also generated specialized schools of thought, but the issues that defined (and divided) them tended to be scholarly rather than political. The most prominent of these American schools was based for many years at the University of California–Berkeley, and was dominated by the powerful personality of the cultural geographer Carl Sauer. The core idea of this school was the notion that a society's lifeways would be imprinted on the Earth as a *cultural landscape* that could be subjected to spatial analysis wherever it was found.

Geographers not only take a wide view, but also a long view. We try not to lose sight of the forest for the trees, and put what we discover in temporal as well as spatial perspective. "Geography is synthesis," is one fairly effective answer to that question about just what geography is. That is, geographers try to find ways to link apparently disparate information to solve unanswered problems. As you will see later, sometimes such daring generalizations can set research off in very fruitful directions.

These days, though, it takes courage to generalize and hypothesize. This, as we all know, is the age of specialization. But specialized research ought to have some link to the big questions that confront us, or you have reason to question its value. Fifty years ago one of my professors at Northwestern University often urged me and my co-students to practice what he called "intelligent dinner conversation" (a quaint cultural tradition, remnants of which are still observable in certain urban settings). "Always be ready to explain in ordinary language to the guest across the table what it is you do and why it matters," he said. Most of us thought that this was not only unnecessary, but also none of "the public's" business. But he was right, and he would enjoy the debate now going on in professional geography, much of which focuses on ways to speak to the general public in plain language about what it is we do.

Specialization in research and teaching occurs at several levels, of course. I have already mentioned that some geographers (fewer than before) still become area specialists or, in another context, regional scientists. Others study urbanization from various spatial standpoints, and their studies range from highly analytical research on land values and rents to speculative assessments of intercity competition. One especially interesting question has to do with efforts to measure the amount of interaction between cities. When two large cities lie fairly close together, say Baltimore and Philadelphia, there will be more interaction (in numerous spheres ranging from telephone calls to road traffic) than when two cities lie much farther apart, for example Denver and Minneapolis. But just how does this level of interaction vary with city size and intercity distance? The answer is embodied by the so-called gravity model, which holds that interaction can be represented by a simple formula: multiply the two urban populations and divide the total by the square of the distance between them. You can use kilometers, miles, or even some other measure of distance, but so long as you are consistent for comparative purposes the model will do a good job of predicting reality. Distance is a powerful deterrent to interaction—geographers call this distance decay—and measuring this factor can be enormously helpful in business and commercial decision making.

Other geographers combine economics and geography, and focus on spatial aspects of economic activities. The rise of the world's new economic giants on the Pacific Rim has kept them busy.

Still others focus on spatial aspects of political behavior. Political scientists tend to focus on institutions, political geographers on political mosaics. Geopolitics, an early subfield of political geography, was hijacked by Nazi ideologues and lost its reputation; but recently, geopolitics has been making a comeback as an arena of serious and objective research. From power relationships to boundary studies, political geography is a fascinating field.

There are literally dozens of fields of specialization in geography, and for students contemplating a career in geography it's a little bit like being in a candy store. Interested in anthropology? Try cultural geography! Biology? There's biogeography! Geology? Don't forget geomorphology, the study of the evolution of landscape. Historical geography is an obviously fruitful alliance between related disciplines. The list of such options is long, and it is still growing. Developments in mapmaking have opened whole new horizons for technically inclined geographers.

Over a lifetime of geographic endeavor, many geographers change specialties, and I'm one of them. I was educated to be a physical geographer, that is, as someone who specializes in landscape study (geomorphology) and related fields. As such, I spent a year in the field in Swaziland, in southern Africa, trying to determine whether a large, wide valley there was a part of the great African Rift Valley system, the likely geographic source of humanity. While I was preparing for this research, however, I met a political geographer named Arthur Moodie, a British scholar who came to Northwestern University as a visiting professor. I took his classes and never forgot them. When I was hired by Michigan State University as a physical geographer, I also continued to read and study political geography. Eventually, I was asked to teach a course in that field, wrote a book and some articles about it, and thus developed a second specialization.

What I didn't realize, at first, was how my background in physical geography would make me a better political geographer. Like geopolitics, environmental determinism had acquired a bad name between the World Wars, and it could be dangerous, professionally, to try to explain political or other social developments as influenced by environmental circumstances. But I knew that, like geopolitics, environmental studies would make a comeback. When they did, I had the background to participate in the debate. That's how, many years later, I was appointed to Georgetown University to teach environmental issues in the School of Foreign Service.

I made only one other foray into a new field, and that was also as pleasant a geographic experience as I've ever had. It all began with a great bottle of wine. A fateful dinner with that bottle of 1955 Chateau Beychevelle so aroused my curiosity that, five years later, I was working on my book entitled *Wine:*

A Geographic Appreciation, was teaching a course called The Geography of Wine at the University of Miami, and saw some of my students enter the wine business armed with a background they often found to be very advantageous. Geography has few limits—and specialization does have its merits.

BUT IS GEOGRAPHY IMPORTANT?

Remember the bumper sticker, popular some years ago, that said "If You Can Read This, Thank a Teacher"? One day I was driving down one of my least favorite highways, Interstate 95 between Fort Lauderdale and Miami in Florida, when a car passed me whose owner had modified that sticker by inserting the word "Map" after "This" and by pasting a piece of road map at the end of the slogan. I didn't need to ask what that owner's profession was. A geography teacher, obviously.

The fact is, a lot of us cannot read maps. Surveys show that huge numbers of otherwise educated people don't know how to use a map effectively. Even simple road maps are beyond many more of us than you might imagine. People who, you would think, deal with maps all the time and therefore know how to get the most out of them—travel professionals—often have trouble with maps. I live about half the year on Cape Cod, and thus have the dubious pleasure of flying into and out of Boston's Logan Airport, about two hours from home. These days flight schedules are not what they used to be, so when someone arranges my trip I always hope that consideration was given to the other airport about two hours from the mid-Cape, Providence. I've learned not to count on it.

Geography's utility certainly made news shortly after the terrible tsunami of December 26, 2004, when the story of a schoolgirl named Tilly Smith made headlines around the world. Tilly was vacationing in Phuket, Thailand with her parents and was on Maikhao Beach when she saw the water suddenly recede into the distance. She remembered what she had just been taught by her geography teacher, Mr. Andrew Kearney at Danes Hill Prep School in Oxshott, south of London: that the deep wave of a tsunami sucks the water off the beach before it returns in a massive wall that inundates the entire shoreline. Tilly alerted her parents and they ran back and forth, warning beachgoers of the danger and urging them to seek shelter on an upper floor of the hotel nearby. About 100 people followed her advice; all survived. Of those who stayed behind, none did. Britain's tabloids declared Tilly to be "The Angel of Phuket," but give some credit to that geography teacher who obviously had the attention of his students.

Okay, you might say. As an everyday tool to make life a bit more predictable and efficient, and as an occasional environmental alert, geography has its uses. But does that make it important in a general sense?

Consider this: a general public not exposed to a good grounding in geography can be duped into believing all kinds of misinformation. Even today, despite the best efforts of the National Geographic Society and its allies, an American student might go from kindergarten through graduate school without ever taking a single course in geography—let alone a fairly complete program. (That's not true in any other developed country, nor in most developing ones. Geography's status is quite different in Britain, Germany, France, and such countries as Brazil, Nigeria, and India.) This means that when a group of scientists decides to scare the beejeebers out of the public by predicting imminent glaciation (as they did in the 1960s) or looming greenhouse warming (the concern of the 2000s), far too many people are insufficiently informed to be able to make sense of the debate and, through their elected representatives, may be persuaded to spend billions of dollars better invested in other causes.

When I talk about this issue on the public-lecture circuit, someone in the audience is likely to challenge my point about the state of geographic knowledge. It may be bad, goes the argument, but don't worry, our leaders know what geography they need to know. They deal with the world at large on a daily basis, and they're sure to be adequately informed and prepared.

Well, maybe, although I wonder about those leaders who come from elite universities that do not offer any geography as part of their undergraduate or graduate curricula. Do you suppose that, if former defense secretary Robert McNamara had been able to take just one course in basic regional or human geography at his alma mater (Harvard), his perspective on Southeast Asia in general and Vietnam in particular might have been different? I would like to think so, but Harvard University has not offered geography to its students for about a half century. The cost to the country may be greater than we can imagine.

As to our leaders knowing the map they must navigate, consider this little incident in President Nixon's Oval Office, as described by another Harvard figure, former secretary of state Henry Kissinger, in his book *Years of Renewal*:

> As part of some U.N. celebration, the Prime Minister of Mauritius had been invited to Washington. Mauritius is a subtropical island located in the Indian Ocean . . . it enjoys plenty of rainfall and a verdant agriculture. Its relations with the United States were excellent. Somehow my staff confused Mauritius with Mauritania, an arid desert state in West Africa that had broken diplomatic relations with us in 1967 as

an act of solidarity with its Muslim brethren in the aftermath of the Middle East War.

This misconception produced an extraordinary dialogue. Coming straight to the point, Nixon suggested that the time had come to restore diplomatic relations between the United States and Mauritius. This, he noted, would permit resumption of American aid, and one of its benefits might be assistance in dry farming, in which, Nixon maintained, the United States had special capabilities. The stunned visitor, who had come on a goodwill mission from a country with, if anything, excessive rainfall, tried to shift to a more promising subject. He enquired whether Nixon was satisfied with the operation of the space tracking station the United States maintained on his island.

Now it was Nixon's turn to be discomfited as he set about frantically writing on his yellow pad. Tearing off a page, he handed me a note that read: "Why the hell do we have a space tracking system in a country with which we do not have diplomatic relations?" (Kissinger, 1999)

So don't be too sure about geographic knowledge in Washington, D.C. It's pretty obvious that we were not well enough acquainted with the physical or cultural geography of Indochina when we blundered (McNamara's word) into the Vietnam War, and I am sure that many of us had doubts about our leaders' knowledge of the regional or human geography of Iraq in the winter of 2003—remember those cheering, grateful crowds that would line the roads? I often cite that old canard about war teaching geography, but in our case we must add a word: belatedly.

Perhaps the most important byproduct of geographic learning, early or belatedly, lies in its role as an antidote to isolationism. Can there be a more crucial objective than this? In our globalizing, ever more interconnected, still-overpopulated, increasingly competitive, and dangerous world, knowledge is power. The more we know about our planet and its fragile natural environments, about other peoples and cultures, political systems and economies, borders and boundaries, attitudes and aspirations, the better prepared we will be for the challenging times ahead.

From this perspective, geography's importance is second to none.

HOW DID IT COME TO THIS?

There's no denying it: for all its putative importance, geography as a school subject and as a university discipline in the United States is, to put it mildly, underrepresented. This wasn't always the case. There was a time when geography was well established as a discipline at Harvard and Yale, when geography was also widely taught in America's schools. During and after the First World War, through the interwar period and again during and after the Second World War, geography was a prominent component of American education. In prewar debates, wartime strategy, and postwar reconstruction, geographers played useful, sometimes crucial roles. Geographers were the first to bring environmental issues to public attention. They knew about foreign cultures and economies. They had experience with the workings of political boundaries. They produced the maps that helped guide United States policies.

In the 1950s and early 1960s, Americans continued to be well versed in geography. American success during the Second World War had drawn our attention to the outside world as perhaps never before. Maps, atlases, and globes sold by the millions. The magazine with geography's name on it, *National Geographic*, saw its subscription grow to unprecedented numbers. University Geography Departments enrolled more students than they could handle. When President John F. Kennedy launched the Peace Corps, geographers and geography students were quickly appointed as trainers and staffers.

But, as so often happens when social engineers get hold of a system that's working well, the wheels came off. Professional educators thought they had a better idea about how to teach geography: rather than educating students in disciplines such as history, government, and geography, they would teach these subjects in combination. That combination was called social studies. The grand design envisioned a mixture that would give students a well-rounded schooling, a kind of civics for the masses, which implied that school teachers would no longer be educated in the disciplines either. They, too, would study social studies.

Prospective teachers from the School of Education had been among my best and most interested students at the University of Miami during the early 1970s. They registered in large numbers in two courses: World Regional Geography, which was an overview of the geography of the wider world, and Environmental Conservation, a course that was years ahead of its time, and to which even the Department of Biology sent its students. But when the social studies agenda took effect, the student teachers stopped coming. They now had other requirements that precluded their registration in geography.

We geographers knew what this would mean and what it would eventually cost the country. The use of, and knowledge of maps would dwindle. Environmental awareness would decline. Our international outlook would erode. Our businesspeople, politicians, and others would find themselves at

a disadvantage in a rapidly shrinking, ever more interconnected—and competitive—world. Many of us wrote anguished letters to government agencies
and elected representatives, to school district leaders and school principals.
Fortunately, many private and parochial schools continued to teach geography. But for public education, the die was cast.

REVERSAL OF FORTUNE

This set of educational circumstances in little more than a decade produced
exactly what geographers had predicted: an evident and worsening national
geographic illiteracy. All of us who were teaching at the time have stories of
students' disorientation, some of them amusing, most of them worrisome.
By its very name, the catch-all social studies rubric excluded the elementary
but crucial physical geography (including basic climatology) topics that had
been part of the high-school geography curriculum. This was the one subject
in which students got an idea of the importance of understanding human-
environment interactions as well as the workings of climate and weather,
and it was a huge loss. When these students got to college and enrolled in
a first-year geography course, they were at an enormous disadvantage: they
simply did not know these basics.

Some university faculties recognized this situation and decided to do
something about it. Georgetown University was one of them, and I saw the
results firsthand while I was on the faculty of Georgetown's School of Foreign Service from 1990 to 1995. Every incoming student was required to take
a course called Map of the Modern World, a one-credit course offered by the
noted political geographer Charles Pirtle. In one semester, students were expected to become familiar not only with the layout of the political world, but
also with general patterns of geopolitical change, general environmental and
climactic conditions, and resource distributions. It was a tall order, but here
is what impressed me most: at the end of their four-year degree program,
Georgetown students are asked to list the course that pushed their knowledge forward more than any other. Map of the Modern World, a freshman
geography course you would think most students had long forgotten, led the
rankings year after year. It was a tribute to Charlie Pirtle, to be sure—but it
also said something about the relevance of geography in the opinion of these
capable students.

Unfortunately the Georgetown remedial model was (and still is) a rarity,
not a commonplace. The geographic illiteracy of entering freshmen lowered
the level of academic discourse in many an introductory class, and faculty
devised various ways of dealing with it. Some professors were, shall we say,

more sensitive to students' problems than others, and occasionally stories leaked out about embarrassing moments in the classroom. One of these stories involved a colleague of mine at the University of Miami who liked to start his class by asking students to identify a number of prominent geographic locations on a blank map of the world's countries. The results were always abysmal, and they grew worse as time went on. The good professor would grade the class as a whole and, reportedly with biting sarcasm, would announce the large percentage of participants who could not locate the Pacific Ocean, the Sahara, Mexico, or China.

Early in the fall semester of 1980, the student newspaper, the *Miami Hurricane*, got hold of the test, a summary of test results, and the professor's witty commentary. The paper's front-page story on this tale of "geographic illiteracy" was picked up by the major news media. NBC's *Today* show appeared on campus. ABC's *Good Morning America* invited the principals to New York, but the segment was too brief to throw real light on the problem.

The news, however, had spread throughout the country, and while officials at the University of Miami fretted about what the story might do to the university's reputation, teachers elsewhere tried their own tests on their students. We are all too familiar with the results. At one Midwestern college, only 5 percent of the students could identify Vietnam on a world map. At another college, only 42 percent correctly named Mexico as our southern neighbor. Specialists, including some of the very educators who had helped engineer the demise of school geography, claimed to be "dismayed" at such results. While geographers were not surprised, the question was: how would we reverse this ignominious tide of ignorance?

ENTER THE SOCIETY

Tales of on-campus geographic blindness soon led to newspaper stories of public illiteracy as well. Journalists took to the streets with outline maps of the United States and of the world, asking people at random to identify such features as New York State and the Pacific Ocean and (so it seemed) gleefully reporting the embarrassing tallies. Their stories, however, were usually buried among marginalia.

But then something happened to change the picture quite radically. President Reagan, upon arriving in Brasilia, the capital of Brazil, to open an important international conference, pronounced himself pleased to be in . . . Bolivia. This caused quite a stir in Brazil, and his faux pas made the front page of *USA Today*, which busied itself identifying similar gaffes by other politicians. Now geographic illiteracy suddenly was headline news, and the television networks fell over themselves covering it. One of them, ABC-TV, called the University of Miami, which relayed the call to me at a hotel in Baltimore where I was attending a meeting. That call led to my first appearance on *Good Morning America*, and the response to my segment (from the Netherlands) generated a week-long geography series a few months later and my six-year appointment to the *GMA* staff as geography editor subsequently.

But it would take more than the support of *GMA*'s perceptive executive producer, Jack Reilly, to make a real dent in our national geographic illiteracy. As it happened, however, I had a parallel opportunity through my appointment as an editor at the National Geographic Society in 1984. In 1980 I had had the good fortune of being invited to join the Society's Committee for Research and Exploration, and I began almost immediately to discuss ways of involving the Society in the campaign. The Society's president, Gilbert M. Grosvenor, was sympathetic to the idea. He seemed to be galvanized by a Society-commissioned Gallup Poll that proved without a doubt that American students had fallen far behind their European and other foreign contemporaries in terms of their geographic knowledge. When I joined the NGS staff full time in 1984 for a six-year editorial term, I was able to help mobilize a crucial alliance.

To most observers, it would have seemed natural for an organization known as the National *Geographic* Society to come to the aid of the discipline. But it was not so simple. For many years, the Society and the discipline had not enjoyed good relations. To the Society and its leadership, professional geographers seemed snobbish, insulated, and often unimaginative. To professional geographers, the Society's popularization of its magazine and the rubric of geography was inappropriate and misleading. "There's precious little geography in *National Geographic*," said my professor at Northwestern University when I arrived there as a graduate student in 1956. "If you're going to subscribe, you'd better have the magazine sent to your home. Not a good idea to see it in your department mailbox here."

That amazed me. In fact, when I was living in Africa during the early 1950s, *National Geographic* was my window to the world, its maps a source of inspiration. I had written its president, Gilbert H. Grosvenor, in 1950 to tell him so. He sent a gracious letter in response, urging me to continue my interest in geography and inviting me to visit the Society's headquarters "if [I] were ever to come to the United States." But as a graduate student, I soon realized that the National Geographic Society and its publications were generally not held in high esteem among "professional" geographers.

Grosvenor's grandson, Gilbert M. Grosvenor, however, was not one to let such bygones get in the way. He launched a massive financial and educational campaign in support of geography at the school level, realizing better than most of us that the schools and their teachers were the key to the future of the discipline. High-school students, he knew, were not coming to college in any numbers intending to major in geography, because they never saw geography as an option when they graduated. The social-studies debacle had pretty well depleted the ranks of geography teachers, so the first order of business was to prepare large numbers of teachers to teach a geography curriculum. Since the geography curriculum itself had atrophied, Grosvenor appointed a prominent specialist in geographic education, Christopher "Kit" Salter, to resurrect it. Salter, in consultation with the half dozen or so geographers on the Society's staff, developed a spatially and environmentally based framework that would come to be known as the "Five Themes" of geography. In 1986 the Society printed several million copies of an annotated map in full color titled "Maps, The Landscape, and Fundamental Themes in Geography," providing every school in the country with as large a supply as needed.

Meanwhile, Salter under NGS auspices organized a nationwide network of so-called Geographic Alliances representing every State in the Union. These alliances consisted of geography teachers supported by the Society in various ways. Representatives of each alliance were invited to Society headquarters in Washington, D.C. for instruction in geographic education; they would in turn assemble teachers in their home States to convey what they had learned. Thus the number of teachers competent to teach geography increased exponentially, as did grass-roots support for the revival of the subject in schools all over the country.

Grosvenor raised significant funding for the project, testified on Capitol Hill on behalf of geography as an essential component of national education standards, buttonholed politicians, and crisscrossed the country speaking for geography. Not everyone on his staff in Washington was enthused by, or even supportive of, all these efforts, and not all professional geographers ensconced in their academic departments appreciated what he did. But the leadership of the Association of American Geographers had the good sense to extend formal recognition to him for a campaign that closed the book on old, painful disharmony between Society and discipline.

So where are we today? I wish I could report that all the foregoing drastically altered the level of exposure of elementary and high-school students to geography. But the road back will be long and hard. The best assessment is that when the Society's campaign began, about 7 percent of American students were getting some geography; today, after nearly 20 years and an estimated investment of $100 million, the figure is still below 30 percent. We have a long way to go.

WILL GEOGRAPHY BE HISTORY?

Some of my colleagues take a dim view of the future of geography as a discipline. Yes, the United States Congress endorsed the establishment of National Geography Week every November, and the winner of the annual National Geography Bee, modeled on the famous spelling bee, gets television coverage every spring. Largely due to the National Geographic Society's efforts, geography was designated as one of the five cornerstones of American education during the tenure of (then) education secretary Lamar Alexander in the first Bush administration (geography did less well in the "No Child Left Behind" program of the second Bush administration).

But against these promising developments in the public arena stand some worrisome negatives, two in particular. Ours is a history-obsessed culture. From archeology to geology and paleontology to linguistics, we tend to focus on the temporal. In higher education, spatial science gets short shrift just as geography still does at the school level. To Americans it is inconceivable that a university or college, whether prestigious or unpretentious, could exist without a history department. No basic curriculum, whether at Harvard or at a Midwest community college, would exclude a history component. The same cannot be said for geography.

And professional geographers, as we have noted, are divided on the substance of their discipline. It's probably a healthy debate, it isn't the first time, and it goes on in other disciplines too. But it can be confusing to college and university administrators who read our scholarly journals and aren't sure just what our consensus is. History, anthropology, and biology are more clearly defined—they think.

I take a fairly Neanderthal view of this issue. Our basic, common ground, I feel, lies in regional geography, human-cultural geography, and physical (environmental) geography, along with the analytical tools students will need as they begin to specialize even at the undergraduate level, ranging from statistical analysis to Geographic Information Systems. Beyond this, the tie that binds us—but need not constrain those who go off in other directions—is the spatial perspective and spatial analysis. To those who doubt geography's disciplinary future I say that our great opportunity lies at the interface of environment and humanity. We have been at this for the better part of a century and we were ahead of our contemporaries for much of that time. We should reclaim our position.

As to geography becoming history, I must tell you that I admire and envy the way historians have made their case to the general public as well as academically. Every time I turn on my television I seem to find some "presidential historian" commenting on good deeds and misdeeds of former presidents. And I agree: it is true that we should be reminded now and then of what President Nixon knew about Watergate and when he knew it. *When*, after all, is history's key question. But more recently we had a president who evoked the question: what did the president do and *where* did he do it? That's geography! We need a presidential geographer! My proposals to this effect have, for some reason, been ignored by the networks.

GEOGRAPHIC LITERACY AND NATIONAL SECURITY

Geographic knowledge is a crucial ingredient of our national security. We have crossed the threshold to a century that will witness massive environmental change, major population shifts, recurrent civilizational conflicts, China's emergence as a geopolitical as well as an economic superpower, unifying Europe's transformation into a major player on the international stage—among other developments yet unforeseen. Among my colleagues are geographers who conduct research on the likelihood of coming energy crises and how to forestall them, on the risks of WMD (Weapons of Mass Destruction) dissemination and how to mitigate them, on the impact of global climate change in especially vulnerable areas and how to confront it. These are serious issues indeed, and while geographic knowledge by itself cannot solve them, they will not be effectively approached without it. WMD diffusion, for example, is driven by technology as well as ideology. The technology is the stuff of other disciplines, but ideology has significant geographic ramifications. Extremism of the kind that propelled the Taliban movement to power in Afghanistan from its bases in mountainous and remote western Pakistan tends to fester in isolated locales, and there is nothing uniquely Islamic about this. States that fail, at dreadful cost to their inhabitants, tend to lie segregated from the mainstreams of global interaction and exchange. From Somalia to Afghanistan, from Cambodia to Liberia, from Myanmar to North Korea, their peoples pay a terrible price.

Geography is a superb antidote to isolationism and provincialism. Some specialists in geographic education argue that our persistent national geographic illiteracy results from our own "splendid" isolation between two oceans and two nations, but we are learning that this spatial solitude means little in a fast-globalizing world. During the Vietnam War, there were politicians who advocated "bombing the North back to the Stone Age," and the United States had the power to do so. What the United States was unable to do was to persuade tens of millions of Vietnamese to change their ideology. More recently in Iraq, military intervention proceeded quickly and efficiently, leading to premature assertions that the war was won. But the real war, for Iraqis' hearts and minds, still lay ahead and entailed a costly insurgency that devastated the country's heartland and whose end is not in sight. The United States and its coalition had equipment and ordinance, but could not prevent the alienation of a growing minority of mostly Sunni citizens. Too few Americans know the region, speak the languages, comprehend the faiths, understand the rhythms of life, realize the depth of feelings. And even as the campaign in Iraq continues, other dangers loom. Among these, the most significant in the long term is the coming contest with China—but how much more does the general public in America know about China today than it (or its leaders) knew about Southeast Asia four decades ago?

If there was a way to mobilize it, I would not only reinstate departments of geography in our "elite" universities but also resurrect regional studies in all such departments, old and new, to ensure that, once again, a growing cadre of field-experienced, language-capable, locally connected scholars would populate government, intelligence, and other national agencies whose efforts will be at least as important as high-altitude weapons delivery, satellite imagery, and GIS scrutiny. Geography, unlike its public image, is an entertaining as well as enlightening field, but what follows is also serious—dead serious.

2

READING MAPS AND FACING THREATS

It is often said that a picture is worth a thousand words. If that is true, then a map is worth a million, and maybe more. Even at just a glance, a map can reveal what no amount of description can. Maps are the language of geography, often the most direct and effective way to convey grand ideas or complex theories. The mother of all maps is the globe, and no household, especially one with school-age children, should be without one. A globe reminds us of the limits of our terrestrial living space when about 70 percent of its surface is water or ice, and much of the land is mapped as mountains or desert. A globe shows us that the shortest distance between the coterminous United States and China is not across the Pacific Ocean but over Alaska and the Bering Sea. A globe tells us why Northern Hemisphere countries dominate the affairs of the world: most habitable territory lies north of the equator.

Cartography, the drawing of maps, has come a long way since ancient Mesopotamians 5,000 years ago scratched grooves in moist clay to represent rivers and fields and let the sun bake it into clay tablets. The evolution of cartography is a stirring story well told by John N. Wilford in *The Mapmakers* (Wilford, 1981), and the saga continues. During the first period of the still-continuing age of discovery, explorers, mercenaries, speculators, and adventurers sailed from Europe into the unknown, and those who survived brought back pieces of the great global puzzle for cartographers to fit into their maps.

Magellan and his crew were the first to circumnavigate the world (1519–1522), building on Cabral's impressions of the coast of Brazil and proving the vastness of the Pacific; the Italian Battista Agnese's 1544 map of the world was soon renowned for its beauty as well as its novelty; and the Flemish mathematician and cartographer Gerardus Mercator formulated a grid for the evolving map of the world that allowed navigators to plot a straight-line compass bearing, the Mercator Projection (1569). This was a momentous innovation, and the name Mercator remains famous to this day—as well as another of his inventions, the concept of the atlas (which he named after a Greek titan) as a collection of maps.

It is well to remember, however, that Europeans were not the only map-makers of the time. The Chinese were making maps perhaps 3,000 years ago, and their fleets, larger than anything Europe had floated, were plying Asian and African coasts before Magellan made his epic journey. When Captain Cook traversed the Pacific in the eighteenth century, the local islanders showed him the way through maps made of sticks, fibers, and shells. The ancient Maya and the Inca also made maps. Mapmaking was not a European monopoly.

But the Europeans did collect and assemble the information necessary to create the first representative maps of the entire world rather than just their own realm, and they got better at it as time went on. From Vasco da Gama at sea to Burton and Speke on land, "discoveries" (a word to which, these days, the descendants of the discoverees tend to object) were made at ground or from water level. The explorers and their improving equipment, ranging from compass to sextant and from astrolabe to chronometer, eventually achieved remarkable accuracy and amazing interpretive detail. Nineteenth-century maps tend to represent science, not art, although many were still hand colored for clarity. But the twentieth century witnessed the revolution that would transform cartography and is still under way: the introduction of photography from airplanes, the launching of image-transmitting orbital satellites, and the coming of the computer age.

In the process, the very definition of the term "map" has changed. In traditional works on cartography such as *The Nature of Maps* by Arthur H. Robinson and Barbara B. Petchenik (1976) or P. C. and J. O. Muehreke's *Map Use* (1997) the authors define a map as "a graphic representation of the milieu" or "any geographical image of environment." But today we see maps of the brain, of human DNA, of ozone holes in the atmosphere, of Mars, of galaxies. High technology, in the words of Stephen S. Hall, "has completely stolen cartography from the purely terrestrial domain" (Hall, 1993).

Well, not entirely. Scientists may use "maps" of chromosomes, regions of the brain that activate when music is heard, or galactic realms where the cosmic action is, but we still use maps for planning a trip, for getting around, for checking on the weather, for getting a sense of where something important in the country or the world is taking place. Unfortunately, surveys show that many Americans are unable to make full use of such traditional maps, even simple ones in commercial road atlases. They have trouble dealing with the standard properties of ordinary maps, such as scale, orientation (direction), and symbols. They find it difficult to relate the legends of maps to the contents of the maps themselves. It is also easy to be confused by the effects of certain map projections, for example the Mercator map, which has the asset of directional utility, but at the cost of shape and size. On a Mercator map, Greenland looks bigger than South America when, in fact, South America is eight times as large as Greenland. Don't plan your overseas trip with a Mercator map!

MAP SCALE

There is no escaping it: a map, if it encompasses a section (or all) of the Earth's surface, must represent a rounded surface on a flat piece of paper. The larger the segment of the globe thus represented, the greater the problem. It hardly matters when it comes to a map of a small town, where the curvature of the Earth has little effect. But a map of the entire United States needs substantial "flattening," and a map of the entire world requires some complicated manipulation to avoid crippling distortion.

The larger the area represented on our flat piece of paper, the smaller the scale of the map and the less the detail it can display. This is one of those apparent contradictions in the language of geography: you would think that a map of a whole continent is a large-scale map, because it covers such a large area, but in fact it is a small-scale map. A page-size map of a city block or suburban street where you live covers a small area, yet it is a large-scale map. At this large scale, you can show individual houses, streets, and sidewalks. Most of this detail would be lost, however, if the page were to contain a map of an entire city. Now the scale would be smaller, and only major urban areas and arteries could be shown. Put an entire state on the page, and the city becomes little more than an irregularly shaped patch. In turn, the state becomes just an outline if the page must contain the whole country, with very little detail possible.

Each time, in the example above, we made the map's scale smaller to accommodate an ever larger area on our page. Thus, in addition to the legend and its symbols, we should examine the scale of any map we read. Our expectations of what the map can tell us are based in part on the scale to which it is drawn.

Why is scale smaller when the area represented becomes larger? Because scale refers to a ratio: the ratio of a distance or area on a map to the actual, real-world distance or area it represents. To simplify, let us use the number one for the distance on the map. On our map of a city block or suburban street, 1 inch would represent about 200 feet, or 2,400 inches, so the scale would be 1:2,400. This ratio can also be represented as a fraction: 1/2,400. To get the whole city on the page, 1 inch would have to represent 2 miles, or 126,720 inches. Now the ratio (1:126,720) becomes a much smaller fraction:

1/126,720. As maps go, 1:126,720 still is a pretty large-scale map. To get an entire medium-sized state on our page, our scale would have to drop to about 64 miles to the inch, or 1:4,000,000. And for the whole United States, the scale would be 1:40,000,000.

The scale of map, therefore, tells us much about its intended use. When a developer lays out a new subdivision or a planner considers the placement of a new shopping center, large-scale maps are needed. The useful road maps made available by the American Automobile Association and by state tourist offices are at medium scales. Maps of world distributions (of, say, population growth by country) can be presented at small scales. The map's function is the key to its scale.

DISTANCE

A good map is likely to display its scale in one of two (sometimes both) ways: as the fraction just discussed, or as a bar graph, usually in kilometers as well as miles. Using this feature apparently is difficult for many map readers, so road maps also show point-to-point distances along highways, for example between exits on interstates. Again, the larger the scale, the more accurate the distances you derive from the bar graph. On a small-scale map of the world, distortion invalidates it for all but the most general impressions.

One way to gauge distances using the bar-graph scale on a map of, say, a country or a state is to mark the distance on the map along the edge of a piece of paper, and then to transfer this to the scale. Recently, following a lecture in Texas, someone asked me how large the "Sunni Triangle" in Iraq is, in square miles. I had a general idea, but when I got back to my office I took my Oxford University Press *Atlas of the World*, marked the distances from Baghdad to Ramadi and from Ramadi to Tikrit, and calculated it (I had estimated 3,000 square miles, and was off by about 200). That bar-graph scale can be very useful indeed.

There is one map on which you can measure distances with complete confidence: a globe. Take a piece of string or tape, and you can measure the shortest distances between places on the planet using the scale provided. That exercise can be interesting as well as troubling. Interesting, because when you lay that string from, say, New York to Beijing or from Los Angeles to Singapore along what geographers call a "great circle" route, that route may not lie where you expected it to. Troubling, because these shortest-distance routes also reveal how close some of our adversaries are to American cities. When the North Koreans began testing rockets and fears rose that such a rocket might deliver a nuclear weapon, strategists on both sides looked at the globe and calculated that Anchorage, Alaska and Honolulu, Hawaii lay just 3,500 and 4,500 miles respectively from North Korean territory. The technology of war is shrinking the protective cushion of distance.

DIRECTION

A third element displayed by a map has to do with orientation. Foreign visitors to the United States often comment on Americans' good sense of direction. Europeans, many of whom come from the old, mazelike cities of their realm, are not used to directions that say "go four blocks east on 23rd Street and three blocks north on Fifth Avenue." They may be walking in an American city's square or rectangular downtown street pattern, but their awareness of north, south, east, or west is not part of their customary personal navigation. In most European cities, compass directions simply aren't useful when it comes to finding the way. Americans, on the other hand, are brought up with reference to the compass rose. Ask about someone's home suburb, and the first reaction often includes a directional reference: "Rosewood lies about six miles west of the city."

But this does not mean that directions cannot be confusing, even to American map readers. It is convention that north lies at the top of a map, south at the bottom, east to the right, and west to the left. Yet some maps are not aligned this way, and even experienced map readers can be confused if this is not made clear by a prominent "north arrow" pointing in some direction other than upward. So in addition to checking the legend's symbols, a map reader needs to check the map's general orientation.

The compass rose consists of more than the four main directions. Midway between north and east the direction is northeast, and between northeast and east it is east-northeast. These refinements are of use mainly in navigation. Which leads us to a useful extension of the compass rose to our wrist watch. Imagine that you are standing on the deck of a boat, headed in a certain direction. That direction may be west-northwest or south-southeast. You spot a school of dolphins, just off to the right of your course, ahead of the bow. Rather than calling out the compass-rose direction, a quicker directional reference would be "Dolphins at one o'clock!" Using the clock so that you are always moving toward twelve o'clock is a great way to share quick information on the highway. "Elk in the meadow at three o'clock!" will have all faces pointed in the proper direction; "Elk over on the right!" is far less specific. When I took students to Africa on safari, we always practiced this method—often with excellent results when the moment of observation was brief.

One final point involving direction. Why is it that the world is always represented in such a way that Europe and Asia and North America are at the top, and Australia and Antarctica are at the bottom? Again, this is a matter of convention. While it is logical to use the Earth's two poles as top and bottom of any world map, nothing in nature specifies that the North Pole should be at the top of the map and the South Pole at the bottom. What is now universal practice developed from the work of the earliest cartographers, who lived in the Northern Hemisphere and who started at the top of their page. Most of what was to be discovered turned out to lie to the east, west, and south of their abodes, and so Africa, South America, and Australia came to occupy the bottom half of the evolving world map. It has been that way ever since, except in Australia and New Zealand. There they pointedly draw maps that put "Down Under" on top of the world.

SYMBOLS

Getting the most out of reading a map involves interpreting the symbols that make many maps look dauntingly complicated. These symbols range from simple dots or circles to mark the location of towns and cities on small-scale maps to terrain representation by means of contour lines on larger-scale maps. The United States Geological Survey (USGS) for a very long time has published a set of Quadrangle sheets in the 7.5 Minute Series covering all of the United States, and if you have not seen the map that covers your home area, you are in for a pleasant surprise, whether you live in a city or in the countryside, in flat or mountainous areas. These USGS maps are not in fact geologic maps, but surface maps: they show in remarkable detail the slopes and streams, roads and paths, forests and lakes, towns and farms, and virtually all else in the natural and cultural landscape. Some even show individual houses. Like a good book, such a map is hard to put down once you have started "reading" it.

USGS and other relatively large-scale maps do require some study of legend and symbol. They show forested areas, generally built-up areas, the location of major electrical overhead transmission lines as well as oil and gas pipelines, power stations and railroad stations, bridges and beaches. Some symbols are obvious, others take some getting used to. But that is the case with all intricate maps that display lots of information.

A major challenge for cartographers is the depiction of hills and valleys, slopes and flatlands collectively called the topography. This can be done in

Fig. 2-1

various ways. One is to create an image of sunlight and shadow so that wrinkles of the topography are alternately lit and shaded, creating a visual representation of the terrain. Another, technically more accurate way is to draw contour lines, as is done on the USGS Quadrangle sheets (Fig. 2-1). A contour line connects all points that lie at the same elevation. A round hill rising above a plain, therefore, would appear on the map as a set of concentric circles, the largest at the base and the smallest near the crest. When the contour lines are bunched closely together, the hill's slope is steep; if they lie farther apart, the slope is gentler. Contour lines can represent scarps, hollows, valleys, and ridges of the local topography. At a glance, they reveal whether the relief in the mapped area (the vertical distance between high and low points) is great or small: a "busy" contour map means lots of high relief.

In the United States, contour lines are still measured in feet; in most of the rest of the world, where metric measures prevail, they are in meters. One reason the United States had been slow in this global conversion has to do not only with long-term cartographic practice, but with something more far reaching: the Ordinance of 1785. According to this decree, land north of the Ohio River and west of Pennsylvania was laid out in a vast township-and-range system based on six-mile "township" squares along parallel, east-west baselines before it was opened to purchase by farmers. Eventually the system allocated most of the land between the Appalachians and the Rockies, creating the rectangular cultural landscape familiar to anyone who has flown over or driven through Midwestern States. For more than two centuries all the titles and other legal documents pertaining to this vast expanse of the United States have been expressed in English measures; converting it all from miles and acres into kilometers and hectares would be impractical. So nonmetric practice continues here, grooved ineradicably into the soil.

MAP PROJECTIONS

This brings us to the interesting topic of map projections. Earlier I mentioned Mercator and his milestone navigation-friendly projection, without saying exactly what a map "projection" is but pointing out that projections inevitably distort the reality on the ground (Greenland's huge size on the Mercator projection is a case in point). The fact is that you can manipulate map projections to exaggerate, diminish, distort, and otherwise modify any representation of any part of the Earth's surface. Countries can be made to look larger, compared to others, than they really are. Places can be made to look closer to you than they really are. Maps can be used for propaganda purposes, or worse.

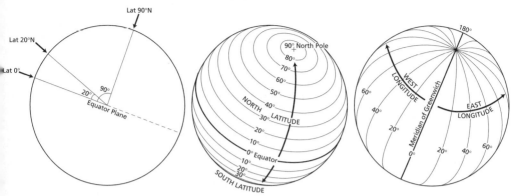

Fig. 2-2

They can be used to stimulate fear, intimidation, aggression, anger, or, at the very least, misjudgments among their readers. So, with any map as with the written word, reader beware!

For centuries mapmakers have grappled with the problem of representing our spherical Earth on a flat surface. To get the job done, they constructed an imaginary grid around the planet, using the poles of rotation and the globe-bisecting equator as their starting points. Since a full circle has 360 degrees, the Earth is divided, pole to pole, by meridians (Fig. 2-2). Of course a starting meridian, or prime meridian, was needed, the zero-degree meridian. This decision was made when the British Empire was at its zenith, and so, not surprisingly, the prime meridian was established as the line of *longitude* running through the Greenwich Observatory near London. That turned out to be a fortunate choice, because the 180-degree line, which would divide the globe into Western and Eastern Hemispheres, lay on the opposite side of the world from London—right in the middle of the Pacific Ocean.

The meridians are the "vertical" lines of the Earth's grid; they converge on the poles and are farthest apart at the equator. The "horizontal" lines, called parallels, meet nowhere: they form equidistant rings around the globe, starting at the equator. These parallels, too, are numbered by degree: the equator, being neither north nor south, is zero degrees *latitude*. Then the numbers go up. Mexico City lies just below 20 degrees north latitude; Madrid, Spain, just above 40; St. Petersburg, Russia, at 60; and Russia's northern islands around 80. The North Pole, of course, lies at 90 degrees, the highest latitudinal point on Earth.

So why is a globe or map grid called a projection? Because that's exactly what it is. You can imagine this by considering an open wire grid with a light bulb at the center and a cylinder of paper wrapped around it. The parallels and meridians would throw shadows onto the paper, creating a kind of

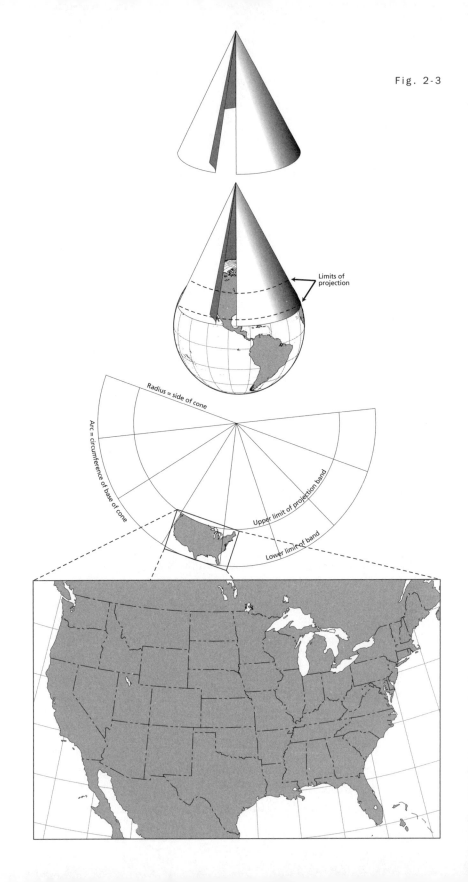

Fig. 2-3

Limits of projection

Radius = side of cone

Arc = circumference of base of cone

Upper limit of projection band

Lower limit of band

projection—not a usable one, but now it becomes a matter of manipulating the light source or the paper. If you put a neon tube inside, pole to pole, you would get something close to the Mercator projection. If you made a conical "hat" of the paper and put it on the Northern Hemisphere, rather than a cylinder all around the globe, you would get less distortion—but only half the world. Indeed, some projections are called "cylindrical" and others "conical." If you want to produce a comparatively low-distortion map of the United States, you would use a conical projection (Fig. 2-3). Projections designed to minimize the distortion of shape are easier to devise for limited latitudinal areas like the coterminous United States than for the world as a whole, but some remarkable, inventive projections have been created nevertheless. By "interrupting" such projections in the oceans, where losing parts of the depicted surface matters less, the shape of the continents is remarkably preserved (see, for example, Fig. 5-2, page 98). Projections designed to keep continents and countries as close to their shape and size as possible are called equal-area projections.

To identify what geographers call the absolute location of a place (as opposed to relative location, about which more later), we must state the degrees, minutes, and seconds north or south latitude and east or west longitude. That seems cumbersome at first, but when you get used to it you can quickly pinpoint the remotest spot on the globe (atlas indexes provide these data for hundreds of thousands of locations). Importantly, no two places have exactly the same location. The recent invention of Global Positioning System (GPS) technology now makes it possible for scientists (for example, archeologists) to record the location of a discovery, leave it, and relocate it even if a sandstorm has changed the terrain unrecognizably. From submerged wrecks at sea to cave entrances on land, from a single dwelling in an isolated village to a gravesite in an overgrown valley, GPS equipment records coordinates that are unique to each. This is one of those "how did we ever do without it" inventions that have radically changed scientific research over the past few decades.

To return to the global grid of parallels and meridians, Figure 2-2 suggests an important reality: latitude and longitude lines cross at right angles. Long before the modern grid was laid out, Mercator realized the implications of this, creating his navigator's chart when much of the world was still to be "discovered." What Mercator probably did not foresee was that his projection, once all lands had been identified and mapped, would be used for other purposes as well (Fig. 2-4). The Russians, Europeans, and Canadians loved it. Leaders and teachers in midlatitude countries liked a map that boosted the size of their homelands.

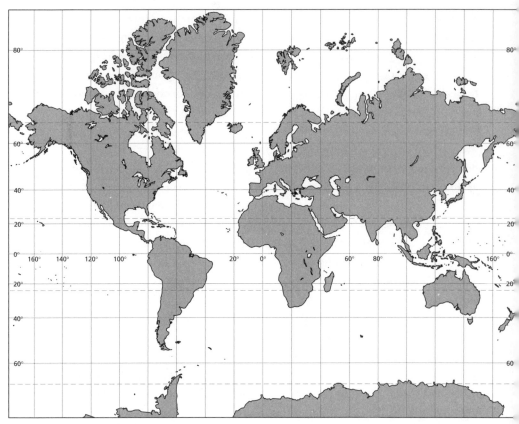

Fig. 2-4

For nearly a century, the National Geographic Society often used the Mercator projection to display world political changes. Then, during the 1980s, the Society's leaders decided to adopt a different projection as their standard world map, a projection constructed by the American geographer Arthur Robinson (a number of other less-distorting projections were also in use at the time). As editor of the Society's scientific journal, called *National Geographic Research*, at the time, I was invited to the news conference where the change from Mercator to Robinson was announced (Fig. 2-5). When it came time for questions, a reporter from a local news organization rose and asked, "Why has it taken you so long to make this change? It seems to me that the old map reduces the size of Africa and other tropical areas, and it's a kind of cartographic imperialism!" She had a point: the Mercator projection gives not only the United States but also the former colonial countries a huge boost in size. But, as the Society's representative said in response, the biggest "loser" from Mercator to Robinson was the Soviet Union. It "shrank" by as much as 47 percent!

MANIPULATING MAPS

By bending the Earth's grid lines in certain ways, therefore, we can make continents or countries look bigger than they really are, or smaller, or shaped differently. Cartographic deception, intended or accidental, is more common than you might think. The misuse of maps has a long history. The Nazis were masters at it. The Communists could match them. Advertisers, politicians of may stripes, activists promoting their causes, and others have used misleading maps in pursuit of their objectives.

Sometimes the misuse of maps is unintended, simply a reflection of our general geographic illiteracy. It was painful to receive last year a copy of a book titled *Middle East for Dummies* with a map of the region on the cover more than 30 years out of date, still showing the diamond-shaped "Neutral Zone" that once lay between Saudi Arabia and Iraq, just west of Kuwait. Saudi Arabia and Iraq agreed in 1973 to establish a boundary down the middle of this historic remnant, eliminating it from the political landscape. But there it was, on the cover of a 2004 book. Dummies indeed (Davis, 2004).

My colleague Mark Monmonier of Syracuse University's Geography Department has written a series of superb books on many aspects of maps, including deliberate deceptions. His first, *How to Lie with Maps*, points out that while some cartographic distortion is actually intended to deceive or misinform, other maps owe their misleading content to sloppy cartography, poor map design, and inexperience on the part of the mapmakers (Monmonier, 1991). You can confirm this by keeping track of maps in the otherwise

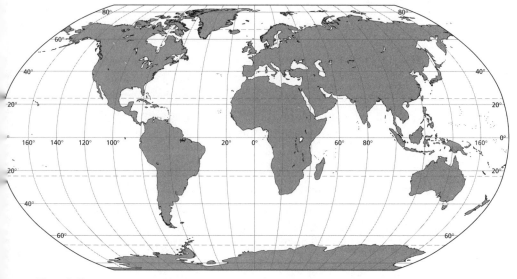

Fig. 2-5

punctilious *New York Times*: chances are that a map on Tuesday will be followed by a correction on Wednesday. The problem, as I found out at ABC and NBC, is that maps are often drawn by people whose background is in art, not cartography. Their maps violate all kinds of rules and conventions because they never had any formal training in a geography department's cartography course. Every school of art or design should make a course in introductory cartography mandatory. Every major newspaper should have someone on its staff with some formal knowledge of mapmaking.

There are encouraging indications that this is beginning to happen. The *New York Times* is publishing more sophisticated, thematic maps, based on extensive independent research, to support stories about major international developments. *USA Today* uses color maps to chronicle social change in America. Occasionally newspapers invite geography graduate students with cartographic skills to apply for positions as interns. All this can only enhance their readers' geographic literacy.

As Monmonier points out, there is power in cartography. Imagine that you are opposed to the building of some installation, say a power station, or a waste dump, or a jail, in or near your neighborhood. You plan to distribute a map at the public meeting where this issue is to be discussed. How to enhance the impact of that map? Obviously, the distance from this new facility to the homes most immediately affected by it is a key factor. You can represent this by drawing concentric circles to reflect the anticipated impact—but the map will have much more effect if you make the lines ever thicker as they approach the NIMBY "victims." Now the map conveys the threat far more effectively.

Monmonier's sharp eye spots another, rather less serious, form of deception—a bit of mischief on the part of a cartographer wielding his or her bit of unchecked power. Scanning a 1979 road map of Michigan and northern Ohio, he noted a couple of hamlets with unusual names: Goblu and Beatosu. Take the road past these places, and you will find that they don't exist. The cartographer, obviously a Wolverine fan, must have been carried away by an approaching Michigan-Ohio State football game. Just don't plan to stop for gas at Goblu or Beatosu.

If you have read about hurricanes, tornadoes, earthquakes, volcanic eruptions, floods, or other natural disasters, or perhaps experienced one or more of these terrifying events, you have probably asked yourself where it is safest to live on this continent. In *Cartographies of Danger*, Monmonier warns us that natural disasters aren't all we should be concerned about (Monmonier, 1997). His maps of hazard zones around nuclear power stations, large incinerators, natural-gas pipelines, and other risk-generating facilities in the cultural landscape remind us that we are also a danger to ourselves, especially when we circumvent rules and regulations designed to protect us.

THE EVER-CHANGING MAP

I have limitless admiration for cartographers. They draw boundaries and put names on maps, go to bed at night, and wake up in the morning to find that boundaries have shifted, new countries have formed, and names have changed. Their job is never done. Prepare an atlas, and it will—especially these days—be out of date before it is published.

What is happening today is not really new. During the age of exploration, new information kept arriving on the drawing tables of European cartographers, and maps had to be revised constantly. But in the period following decolonization and before the collapse of the Soviet empire, there was a certain stability, and map changes were—comparatively—few.

Then, during the 1990s, cartographers faced a flood of new names for cities, towns, and other features in the former Soviet Union, followed by another surge of changes emanating from collapsing Yugoslavia. This came hard on the heels of China's momentous transition from the old spelling to the Pinyin system. I had a foretaste of this when the Society's Research Committee, on which I served, made a month-long visit to China in 1981. It took some getting used to, reading Beijing for Peking, Xizang for Tibet, Guangzhou for Canton. A colleague in Guangzhou smiled when he asked what we Americans were going to do now, familiar as we were with "Cantonese cuisine." Guangzhouese cooking? It would never do! Just don't change the name of Sichuan, we said. At least we can still pronounce it, and we love the taste.

The changes in the Soviet Union produced some similar conundrums. When the Ukraine became independent, its leaders changed the name of the famous old capital, Kiev, to Kyyiv. Chicken Kyyiv?

When the National Geographic Society published the revised version of the sixth edition of its *Atlas*, long known for its place-name coverage, it contained approximately 10,000 changes, yet it was out of date when it hit the streets. The veritable flood of new names continued through the end of the decade, and into the new century. South Africa delimited nine new provinces. India added three new states. But some of the new names were not generally approved, for example the Turkish Republic of Northern Cyprus, recognized only by the Turks, and the Republic of Somaliland, a breakaway province of Somalia. Ever heard of a country called Puntland? That's still another part of Somalia where the people, eager to stay out of the conflicts and chaos that prevail in the south (where the capital is located), proclaimed autonomy.

Cartographers not only add or change names, they must also decide what they will and will not put on their maps, and their decisions can produce great controversy. When Yugoslavia collapsed and its "republics" became sovereign states, the Greeks objected vehemently to one of them: Macedonia. The Greeks argued that the name of one of their historic provinces could not simply be appropriated by a neighboring country. The Macedonians said that they had lived in a political entity of that name for generations. The Greeks proposed that Macedonia be called Former Yugoslav Republic Of Macedonia (FYROM). The Macedonians called that ridiculous. So what's a cartographer to do? Whatever goes on the map will stir up anger somewhere. Cartographers cannot please everybody.

My colleague Ki-Suk Lee of Seoul National University reminded me of this when he told me that my use of the name Sea of Japan for the waters between the Korean Peninsula and the islands of Japan was unacceptable to Koreans. For centuries before there was a Japan at all, he wrote me, Koreans had been referring to this body of water as the East Sea, and at a professional meeting soon thereafter he presented an impressive research paper to support his case. At least I was able to compromise in this instance, which is not always possible. My books (including this one) give it both names, East Sea first, Sea of Japan below between brackets. My books, by the way, sell better in South Korea than they do in Japan.

Trying to be even-handed in situations like this can get you into a lot of trouble. Just ask the people at National Geographic Maps, publishers of the eighth edition of the National Geographic *Atlas of the World* (2005). In every previous edition, the water body between Iran and the Arabian Peninsula had been mapped as the Persian Gulf. After the appearance of the seventh edition, however, a professor from a university in the United Arab Emirates wrote to the Society, arguing that this water body is referred to as the Arabian Gulf in his part of the region, and demanded that the next edition of the *Atlas* show it as the Arabian Gulf, not the Persian Gulf. After much deliberation, it was decided to use both names, with Persian Gulf above and Arabian Gulf, in brackets, below. When the *Atlas* was published, the reaction in Iran and among Iranians everywhere was outrage. Tehran prohibited National Geographic researchers or photographers from entering the country and Iranians in the United States launched a wave of criticism— probably the first time in decades that Iranian rulers in Iran and Iranian exiles abroad agreed on anything. And this campaign had serious impact: on the Amazon.com Web site, it lowered the "rating" of the *Atlas* from five points to two, and anyone unfamiliar with the background might assume that the *Atlas* was not worth its cost, with serious financial consequences for the publisher of this excellent volume. What will National Geographic Maps do in the ninth edition?

Obviously, cartographers are not likely to be out of work anytime soon. We should remember that while mapmaking has entered a new, high-technology age in the United States and in the West generally, maps are still drawn the old way in most of the world—even in the "developed" world. The maps in an atlas I recently edited, the new *Atlas of North America* published in 2005 by Oxford University Press, were drawn in England by cartographers who combined traditional and high-tech methods. Maps in geography (and other) textbooks begin as sketches on paper that are transformed into highly accurate renditions to be checked and revised, and then transmitted electronically to the publisher. Cartography still requires familiarity with design, color gradation, symbolization, and other conventions. Getting the basics right is not always easy.

THE BOARD OF GEOGRAPHIC NAMES

Who decides when the United States, through its official publications, will accept and adopt a new name or a different spelling? This is no minor matter. Our official adoption of a name proposed by one country may profoundly upset another. American approval of the new spelling of a name may make citizens of the place in question very unhappy. Geographic names can be sensitive issues.

Fortunately, the process of approval is done with the utmost care and consideration by an official committee whose members represent a wide range of expertise and experience. This is the United States Board of Geographic Names, an interagency committee that consists of nine members, each representing a branch of the United States government concerned with or affected by such issues. This nine-member board is divided into two standing committees: the Committee on Foreign Names (four members) and the Committee on Domestic Names (five members).

The Committee on Foreign Names consists of representatives from the Defense Department, the State Department, the Central Intelligence Agency, and the Library of Congress. This group considers primarily changes in, and spelling of, country names, important internal divisions (such as the "republics" inside Russia), and international features—that is, features that extend from one country into another (such as mountain ranges and rivers) and whose names may differ on opposite sides of borders. A small army of

staff researchers keeps the committee abreast of changes made within countries, for example, from Leningrad to St. Petersburg, or new spellings, such
as Kyyiv, which are the prerogative of the countries involved.

How does a name change or spelling change become officially approved
by the United States? There are various routes. A government may send a formal communication to the United States secretary of state or to the United
States Embassy in the country, requesting affirmation of a change. If consideration of the change falls within the jurisdiction of the Committee on
Foreign Names, the file is considered there and either approved or deferred,
pending additional information. Eventually, the approved name is codified
during a quarterly meeting of the full Board.

Many names, indeed thousands of them, are changed without such request for approval. Still, the United States needs a consistent form of these
for official use. Thus it is the job of the staff researchers to comb government
decrees, gazettes, maps, and other sources to secure the necessary information. This information is then collected by the Defense Mapping Agency and
submitted to the full Board.

To disseminate the approved names, the Board publishes the *Foreign
Names Information Bulletin*, used throughout the government and elsewhere
(including atlas makers and map companies) for information on accepted
spellings. Only very rarely are the names published in the *Bulletin* not immediately adopted; for cartographers, this is the ultimate source.

MENTAL MAPS

I'm sure you've noticed it: some people seem to have an innate sense of direction. They can find a street or a store with the greatest of ease. They don't
miss highway exits and always know where the one-way streets are.

Others are not so lucky. They get in the wrong lane, can't remember on
which side of the stadium they parked their cars, lose their way trying to find
the home of a dinner host.

Geographers' research has proved that when it comes to the maps in our
minds, our *mental maps*, we are not born equal. Just as some people are color
blind and others have perfect pitch, the brain's capacity to imagine our activity spatially varies from person to person.

I have some evidence of this in my files. Throughout my nearly 50 years of
teaching, I have asked students—not only from America, but from all over
the world—to draw maps of their home city, their state, or their country
(and sometimes the world) on a blank piece of paper. You would be amazed
at the results. Some students can draw, from memory, a remarkably accurate

map not only of their city or state, but of any part of the world. Others cannot even draw the barest outlines. Not all of this is a matter of education or exposure to geography. Indications are that our capacity for what the technical people call "spatial cognition" varies quite widely.

Perhaps you have seen those funny postcards they sell in Texas, showing a map of the United States almost completely occupied by the Lone Star State, with all the others squeezed in a narrow band against the coasts and the Canadian border. Well, I have seen this sort of thing in real life—and not as a joke. During the 1960s, when I taught at one of my favorite institutions, Michigan State University, hundreds of students came from Africa to study there. Quite a few of them took my introductory geography class. I always asked my students to draw their mental map of a continent.

Almost always, the American students drew North America, and the African students drew Africa. And almost always I was impressed by the detail African students put on their maps of Africa. But class after class, year after year, I noted something interesting: many Nigerian students drew Africa the way those Texas postcards show the United States. Nigeria would occupy almost all of West Africa and much of Central Africa, too, and the other African counties would lie squashed around Nigeria's perimeter. Nigeria, to be sure, is a large country, and in fact it is the most populous country in all of Africa. But in the minds of quite a few Nigerian students, it was also the Texas of Africa. "Well, Nigerians think big," said a student from Ibadan when I asked him about this. "Let me try drawing the map again." He did, now mindful of his earlier exaggeration. But when he was finished, Nigeria still was about twice the size it should be.

Mental maps can be improved, of course, through the study of geography. My colleague Thomas Saarinen of the University of Arizona tested students' mental maps throughout the world, with fascinating results. When asked to draw a world map, for example, many students put Europe in the center of it, even when they don't live in Europe and aren't Europeans themselves. That is just one leftover of the educational systems spread worldwide in colonial times.

In general (this should not surprise us), we Americans are rather fuzzier, mental-map-wise, than many of our contemporaries. Does it matter? Geographers think so. As for me, I still remember that day in 1962, when, as a young assistant professor at MSU, I had been invited to join a group of colleagues in the State Department to discuss, with an assistant to Secretary of State for African Affairs G. Mennen Williams, a set of urgent African concerns. But the night before, President Kennedy had appeared on television with a map of Indochina, and in our Washington hotel rooms we had watched his "chalk talk" that revealed the discovery of the Ho Chi Minh Trail. The next morning at State, nobody wanted to talk about Africa. The hot issue was Indochina, and the Trail—and Laos, through which part of it lay. There was much arm waving, numerous proposals and suggestions, but the maps on the wall were of Africa, not Southeast Asia.

"May I just ask," I said as the debate swirled, "can anyone here name the six countries that border Laos? It seems to me that the layout of the region is rather important, given all these ideas."

No one could do it, and worse, no one seemed to care. "That's a waste of time," said one of my colleagues. "If we need to know that, we'll get a map and look at it."

I suggested that if you have a mistaken mental map of a place, you won't know where you're going and that if a whole cadre of decision makers had a vague mental map, we'd be in for bigger trouble than the president had intimated.

In the years since that little incident I've often thought of it and of how little we Americans, on average, really knew about the Indochina where we would wage so costly and bloody a war. Because a mental map is more than a skeleton outline. At its best, it is a store of information, not only about the layout of a place, but also about its components, its dwellings and schools, streets and paths, mosques and markets. It is a map that accrues over a lifetime, the spatial equivalent of our temporal, chronological knowledge and our simultaneous ability to place major events in historical perspective. But our mental maps are not historic, they are current, and they guide and inform our actions and decisions whether a simple excursion for pleasure or a military campaign in a distant land. How clear were they when the Iraq invasion was planned? How clear are they in this century of environmental change, China's rise, and terrorism's threat?

USES AND MISUSES OF MAPS

When geographers are asked to provide an example of the practical utility of maps in solving real-world problems, we like to go back to the story of Dr. John Snow, a London physician-geographer who lived through several of the dreadful cholera pandemics that ravaged much of the world during the nineteenth century. No one knew for sure how cholera spread, making the disease especially frightening, and many victims died within a week of infection. Dr. Snow had come to believe that contaminated water was to blame, but he had no proof of it. When the pandemic that had begun elsewhere in 1842 reached England, London's densely populated Soho District, near Picadilly Circus,

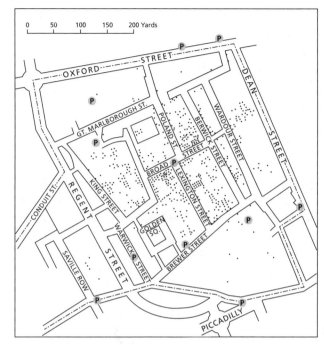

Fig. 2-6

was hard hit. Dr. Snow and his students made a large-scale street map of the area, and marked with a dot the place where each death occurred and where each new case was reported. By the time more than 500 people had died in Soho alone, the map, dated 1854, showed a concentration of victims around an intersection on Broad Street (Fig. 2-6). Now he invoked some geographic reasoning. There were several stores at that intersection, but customers might walk elsewhere if prices a few hundred yards away were lower. But when something is free, they will make a beeline for it, and what was free at this intersection was the water provided by the Broad Street pump. This accounted for the clustering of the dots around the pump, and the link between contaminated water and cholera was confirmed by the map.

That was not the end of the story. Dr. Snow asked city officials to remove the handle from the pump, but they first demurred, saying that this would risk a riot in Soho where people were already angry about the casualty toll from cholera. So Dr. Snow and his students did it themselves, pouring lye down the hole for good measure. Soon he had his proof: the number of deaths around the intersection plummeted, new cases dropped even more sharply, and what the map had confirmed was proven beyond a doubt. Now the authorities could advise people to boil their water and to stop worrying about touching each other or inhaling "bad air," two of the suspected causes.

Today, things are not that simple, but maps still help modern epidemiologists trace pandemics and predict future routes of diffusion. Fast worldwide travel by jet aircraft can relocate a group of individuals infected by some virus in a matter of hours, and once they have dispersed from the airport, warning them before they make contact with locals is a lost cause. On the other hand, cumulative maps of the incidence of such current maladies as Lyme disease and West Nile fever help alert potential victims to the locations of maximum risk even without pointing to remedies.

Against such constructive uses of maps stand the misuses to which they are put. I had a telling experience with this in 1990. A colleague at the University of Baghdad had sent me via a brother in Denmark an official map of Iraq showing Kuwait as the country's nineteenth province. In late July, relations between Iraq and Kuwait deteriorated rapidly, the Iraqis complaining about illegal oil wells along their joint border and massing troops in the area. By coincidence, Dante Fascell, then chair of the House Foreign Affairs Committee, had been invited to the National Geographic Society to speak to the Geographic Alliance teachers who were in Washington attending their institute. Afterward I asked Congressman Fascell about the implications of the map, the troop movements, and the oil issue. He told me not to worry. Our ambassador was on top of things, this was just a quarrel between neighbors, and the United States should not be taking sides. A few days later I happened to be in a hotel in New York for a *Good Morning America* appearance about Puerto Rican immigration the following morning, when the phone rang at two in the morning. It was my producer. "Forget about Puerto Rico," she said. "We've got a car on its way to pick you up. The Iraqis are invading Kuwait. Ever heard of Bubayan Island? Rumaylah Oil Field? We need maps. Start writing the script in the car on the way over here." By the time we sat in the conference room, I had a rough map of the border area drawn on a yellow pad, but I wished I had that map showing Kuwait as Iraq's nineteenth province. When governments start issuing maps that include the territory of neighbors, get ready. They're committing cartographic aggression, in part to see whether anyone is paying attention. Obviously that map raised no eyebrows in Washington. This may have been mistaken for a green light. Fifteen years later, the aftermath goes on.

Cartographic aggression takes several forms, some overt, as in the case of Iraq, others more subtle. In 1993 I received a book titled *Physical Geography of China*, written by Zhao Sonqiao, published in 1986 in Beijing. On the frontispiece is a map of China. But that map, to the trained eye, looks a bit strange. Why? Because in the south, it takes from India virtually all of the Indian State of Arunachal Pradesh, plus a piece of the State of Assam. Now this book in not a political geography of China, nor is the matter of appropriated Indian territory ever discussed in it. China's border is simply assumed to lie deep inside India, and the mountains and valleys thus claimed are discussed as though they are routinely a part of China. Make no mistake: such a map could not, in the 1980s at least, have been published without official approval. It should put not just India but the whole international community on notice of a latent trouble spot.

TRACING THE MAPS OF AGGRESSION

In October 1996 the *New York Times* columnist Thomas Friedman wrote a column under the headline "Your Mission, Should You Accept It" that surveyed the impact on American global leadership and interests of continued cuts in the foreign affairs budget, urging President Clinton to reverse the tide through negotiations with the 105th Congress. A small diversion of military appropriation to foreign affairs, Friedman implied, could yield huge dividends.

Friedman's column reminded me of the results of earlier budget cuts, ones that directly affected geographers—and national security. When I was teaching at Michigan State in the 1960s, one of the foreign-service positions my students could apply for was that of geographic attaché, an embassy or consular assignment whose responsibilities included the monitoring and acquisition of all maps, published officially and unofficially, in the country of record. Their tasks included analysis of the maps they acquired and assessments of their significance. "Such maps," I wrote in a lengthy response the *New York Times* published, along with a drawing by Macieck Albrecht, on October 28, 1996, "provide insights into internal problems and external intentions, the latter often an early warning of aggression." But the position of geographic attaché was eliminated during earlier rounds of budget cuts, and what Friedman was complaining about was only the latest cycle of closure of embassies, consulates, and United States Information Agency libaries. So the question arises: even as countries the world over continue to publish maps that should alert our diplomatic and intelligence agencies to risks to peace and stability, who is now monitoring them? Apparently the lesson of August 1990 went unheeded.

Keeping track of the maps being published by governments of countries or, for that matter, by the governments of States, provinces, or other subunits (I have seen some tourist maps from regional governments that clearly make political points) is only one dimension of a larger task. The general public tends to equate "intelligence" with spying, but much critical information gathered by intelligence agencies comes from sources as mundane as local newspapers, magazines, pamphlets, transcripts, and other nonsecret publications. The problem is that these publications appear in languages and scripts requiring translation and interpretation, and all indications are that the number of foreign-language experts available is dwindling quite rapidly. Fluency in one foreign language and reading ability in another was a standard requirement, rigorously tested, when I was a geography graduate student in the 1950s. By the late 1960s, this had been reduced to one language plus ability in quantitative methods of data analysis. By the late 1970s, the language requirement had been dropped entirely in most graduate schools. Meanwhile, the number of graduate students heading for field research in foreign areas was dwindling as well. It became possible to write doctoral dissertations on, say, internal migration in India or agricultural policy in Japan from generally available information (census data or other government reports, for example) without ever setting foot in India or Japan or being able to converse in Hindi or Japanese. Geography was not alone in this trend, and the cumulative effect was to reduce the level of interaction and familiarity between young American scholars and non-Western colleagues and cultures. There is nothing like living for a substantial time in Mombasa or Chennai or Quito or Tunis, navigating the markets and bazaars, plying the bookshops and reading the local press, interacting with locals while gathering not only data but also what used to be called "field experience." When a government signals its priorities by cutting back on language-capable embassy staffs and cultural-information programs, the effects go far beyond a loss of cartographic inventory.

SENSING REMOTELY

When the Soviets launched Sputnik a half century ago they shook the scientific as well as the military world. Remote sensing, hitherto confined to aerial imagery (infrared photography had been the ultimate technology) expanded to incorporate new instruments that could record a much wider range of the electromagnetic spectrum, that is, the electric and magnetic energy emitted at various wavelengths by all objects on the Earth's surface. An ever-larger number of artificial satellites carrying ever-more sophisticated remote-sensing instruments orbited the planet, communicating directly with computers to create visual images never seen before. The opening pages of modern atlases carry satellite images unmatched by Earthbound cartography, dramatic depictions of topography and river systems, coastlines and glaciers, weather systems and ocean currents.

Among the most productive of these Earth-orbiting satellites has been the Geostationary Operational Environment Satellite (GOES) belonging to the

National Oceanographic and Atmospheric Administration (NOAA). It became possible to place a satellite in a geosynchronous (fixed) orbit, so that it remains stationary above the same point on the Earth's surface. From there, GOES was able to observe the oceans and coasts of the United States, watching for storms and tracking them when they occur. Today, television weather forecasters (some of whom were first trained as geographers) depend heavily on such data. On their televised maps, we can watch weather systems move across the country, a form of animated cartography unheard of just a few decades ago.

Also vital have been four LANDSAT satellites launched between 1972 and 1982 to provide a stream of data about the Earth and its resources. Using a battery of state-of-the-art multispectral scanners and special television cameras, LANDSAT's sensors provided new insights into geologic structures, the expansion of deserts, the shrinking of tropical forests, and even the growth and contraction of algae and other organisms crucial to food chains in the oceans. These satellites enabled the monitoring of world agriculture, forestry, ocean pollution, and a host of other environment-related human activities.

As the capacities of satellite instrumentation improved to the point that it became possible to discern objects as small as cars (and tanks) on the Earth's surface, some satellites were inevitably referred to as "spy" satellites, capable of doing what had required ground reconnaissance before. But countries with a modicum of power and influence were obviously unwilling to have foreign geosynchronous satellites hanging over their territories, so old-fashioned spying continued, as Americans were reminded in 2001 when China forced down a propeller-driven American "surveillance" plane flying just outside its territorial sea, briefly holding the crew and staff hostage on the island of Hainan and causing an international dispute. "Spy, but verify," as Ronald Reagan might have said.

MAPPING SYSTEMICALLY

Coupled with the unprecedented imagery generated by satellite-borne equipment is the equally unparalleled growth of computer versatility, including their graphic performance. Today, the map you see in your favorite magazine may well be drawn by a computer that has been instructed to manipulate information on boundaries, resources, ethnic homelands, or any other spatial feature. That information comes from a Geographic Information System (GIS). In recent years, geospatial technology has added a crucial dimension to geography.

A GIS system is essentially a collection of computers and programs that combine to collect, record, store, retrieve, analyze, manipulate, and display spatial information on a screen; of course the display can be printed out, so that it can be reproduced in atlases and journals. Because of the huge capacity of today's computers, the amount of information they hold is almost infinite. They allow you to select any area of the world and bring it to the screen at whatever scale you desire. A map of the remaining forest cover in Poland? A map of the oil reserves in Angola? The names and locations of India's new States? It is all there for the asking. This is revolutionary change indeed, but there is more. A GIS allows for a dialogue between the map and the map user; no longer is the map static and unchangeable except through laborious alterations in the subsequent editions of published books and articles. The map user asks for information; the computer guides the user toward answers. This is called interactive mapping, and it is the cornerstone of this latest revolution in cartography. Already, some automobiles carry navigation systems that guide the driver toward any address desired, in some cases providing voice as well as graphic instructions for the operator to follow. The applications of GIS are limitless.

In the process, as a recent article in the journal *Nature* asserted, "geospatial technologies have changed the face of geography ... by combining layers of spatially referenced data with remotely sensed aerial or satellite images, high-tech geographers have turned computer mapping into a powerful decision making tool" (Gewin, 2004). The United States government is taking notice: in 2004 the Department of Labor identified geotechnology as one of the three most important emerging and evolving fields, along with nanotechnology and biotechnology. As a result, the job market for geographers is changing as well. "The demand for geospatial skills is growing worldwide, but the job prospects reflect a country's geography, mapping history and even political agenda." In the United States, the focus on homeland security has been one of the many factors driving the job market but, as Gewin points out, there is little demand as yet for remote-sensing expertise in the comparatively small and intensively mapped United Kingdom.

Quoted extensively in this important commentary is the executive director of the Association of American Geographers, Douglas Richardson, who cautions that "although technical skills are important ... employees need a deep understanding of underlying geographic concepts. It's a mistake to think that these technologies require only technician-oriented functions." This is a crucial observation at a time when the burgeoning of GIS training and development would appear to deplete still more the pool of geography graduates who have research, cultural, and linguistic experience in the field. When concern with United States homeland security focuses on GIS tech-

nology to identify, follow, and apprehend terrorist suspects already in the
country, we should remember that the roots of Islamic terrorism lie abroad,
in the streets, souks, bazaars, and mosques of nations from Morocco to Malaysia where local radicals and American researchers once made mutually
instructive contact. The great value of geographers' leadership in GIS should
not be undermined by further withdrawal from the real world beyond the
computer screen's reach.

FACING CHALLENGES

Is geospatial technology an eventual antidote to Americans' still-endemic
geographic illiteracy? Probably not, because the practical side of it will remain mostly confined to domestic use. Getting navigation help to drive to a
restaurant or following a storm via satellite on the Weather Channel will not
significantly broaden horizons or mitigate isolationism. Much was made,
during the quick overthrow of Saddam by United States armed forces in Iraq
in 2003, of the role of GIS in the conduct of the campaign, from instant regional and urban cartographics to supply-line logistics. But GIS wasn't of
much help when it came to the occupation phase. The geographical concept
of a "Sunni Triangle" did not appear on anybody's computer screen until the
tough realities of post-Saddam Iraq set in. Those United States government
officials who predicted that the roadside public would greet American forces
with flowers and gratitude should have unplugged their computers and spent
some time in the field. In the wake of the intervention, Iraq became a destabilized haven for the very terrorists "Homeland Security" wants to constrain. Did those GIS maps show just where Iraq is situated, in the middle of
the Middle East, when the cost of failure was incalculable?

In this opening decade of the twenty-first century, the United States, as
the world's sole superpower, faces challenges and threats near and far. Three
of these challenges are immediate or near term; others are in the wings and
will emerge later. All will be met more effectively if the American public
accepts a responsibility that comes with world leadership: to be better informed about the planet geographically. An informed public, able to express
its views to its representatives as Americans in this democracy are, must
come to play a stronger role in the affairs of state, especially foreign affairs.
After allowing Robert McNamara to steer the country into the morass of
Vietnam on the basis of a perception that he knew more than we did, many
Americans believed that such a thing could not happen again, that the lesson
of Vietnam had been learned. McNamara may have known more than most
Americans about Indochina, but he did not know enough. When it came

to George W. Bush's initiative toward Iraq, the American public was led to
believe that intelligence had uncovered incontrovertible evidence requiring
armed intervention. But did the public know any more about Iraq in 2002
than it did about Vietnam in 1962?

From my geographic perspective the three key challenges the United
States faces in the quarter century ahead are (1) accelerating climate change,
not subject to substantial mitigation through human intervention and therefore requiring national, coordinated preparation; (2) the rise of China as a
regional, then a global force, creating the preconditions for the world's first
intercultural cold war and requiring a reassessment of the United States' role
and objectives in the Western Pacific, and (3) an intensification of the ex-
tremist-Islamic terrorist campaign whose widening circle first touched the
United States in the 1990s and whose occasional successes cannot be allowed
to debilitate the state. These three threats are all understandable in geographic context. Climate change has been a challenge to humanity as long as
humans have existed on this planet, and severe environmental swings have
occurred (without anthropogenic causes) quite recently. The fast-changing
geography of Europe at the start of the severe cooling known as the Little Ice
Age gives us a hint of what may be in store—except that the planet's population is now headed for 7 billion, not 1 billion. With reference to China, there
is a serious asymmetry between what we Americans know about China and
what Chinese tend to know about us. I am not sure that we are better informed about China today than we were about Vietnam in the early 1960s,
with troubling implications. Whenever I am in China, whatever college or
university I visit seems able to round up at short notice a substantial number of interested students whose English is not only good enough to follow
my rapid-fire lecture, but who are able to ask perceptive, often tendentious
questions. How many United States universities could find a few hundred
American Chinese speakers on campus to hear a visitor speak in his or her
native language? And as far as terrorism is concerned, we obviously need a
better mental map of the Islamic world than we have. When Ronald Reagan
was asked, following the disastrous terrorist attack on the marines near Bei-
rut in 1992, why the United States had so many troops in Lebanon, he answered "we're there because of the oil" (Clarke, 2004). But there was no oil in
Lebanon. Oil certainly is entwined with the terrorist threat, as is the Israeli-
Palestinian conflict. However surveys show that many Americans tend to
generalize about the Islamic realm in ways that obscure its diversity, variety,
economic contrasts, and cultural complexity. In this diversity lies opportunity, and geography is the avenue toward understanding it.

Among currently less dangerous but nevertheless serious secondary challenges inevitably involving the United States are the social, economic, and political problems afflicting Subsaharan Africa, the still-uncertain prospects of Russia, the rise of India from regional power to global factor, and the emergence of a European superpower on the economic if not the political stage. All this should be seen against a backdrop of continued population growth, increasing competition for energy resources, and potential nuclear proliferation. It will take unequaled leadership and uncommon public awareness to ensure the stability of the world.

3

EARTH'S CHANGEABLE ENVIRONMENTS

Few topics have aroused as much public debate and dispute over the past quarter century as global warming. It has become more than an argument over information such as temperature readings and evidence such as melting mountain glaciers and thinning polar ice. The global-warming issue pits scientists against politicians, environmentalists against energy-company representatives. The public is bombarded with dire warnings of rising sealevels, disastrous hurricane seasons, torrid summers, and searing droughts. It is risky for scientists to express even the slightest doubt that all of the past 30 years' global environmental change is due to humanity's pollution of the atmosphere. Global warming has become a moral as well as a scientific arena.

It is not surprising that many people do not know whom to believe. If large percentages of Americans cannot identify major physical or political features on a blank map, even fewer could be expected to be able to outline the reasons why atmospheric pressure systems form and move the way they can or ocean currents flow they way they do. An introductory college course in physical geography marvelously summarizes the essentials of the global-warming controversy, because it deals with all the interacting mechanisms that comprise the planetary system, from the evolution of continents to the impact of ice ages and from climate change to biogeography.

What we read in the press is not always helpful. A respected British journal, the *Economist*, not long ago stated in one of its editorial "leaders" that "climate change will be with us for at least another century" after anthropogenic emissions into the atmosphere are brought under control (in one of my numerous unpublished letters to editors, I wrote that after nearly 4.6 billion years of it, we should all rejoice). Another serious newspaper editorialized that "if we can get automobile exhaust gases under control, we can banish global warming forever." With that sort of nonsense from usually reliable sources, it is not surprising that many readers are mystified. Make no mistake: even if all emanations, from factory smokestacks, automobile exhausts, and methane-farting cows stopped tomorrow, global warming would eventually slow, but it would not stop until the natural cycle that drives it goes into reverse. And when that happens we will have a bigger problem still, for global cooling is a far greater long-range threat than global warming at a time when humanity's numbers approach 7 billion on this small planet.

So let us put the global-warming issue in geographic perspective, and take a chronological and spatial look at how we got here, environmentally speaking.

But before we get started, we should—just for this chapter—get used to thinking in terms of millions and billions of years, which isn't easy. One way to go about this is to relate our planet's age to our own. If you happen to be in your midforties, it is easy: a year in your life represents 100 million years of Earthly history. One month equals about 8.3 million years, and a week, just under two. A single day in your life equals some 275,000 years, and one hour about 11,000. Consider this: the emergence of modern humans has taken place, comparatively, in the last day of your life; the rise of modern civilizations, during just the past hour.

If you are in your early twenties, just double these figures; if you are in or near your late sixties, subtract about one-third. No matter what one's age, though, the recency and brevity of our human ascent and domination of this planet cannot but impress. No matter what your age, the dinosaurs held sway until less than a year ago!

DRAMATIC BEGINNINGS

Some 4,600 million years ago planet Earth congealed from an orbiting band of cosmic matter into a fiery ball of molten substance burning fiercely and emitting clouds of superheated gases that found its place in the evolving solar system as the third planet among the nine revolving around the Sun. Millions of lighting bolts struck the red-hot surface while, inside the young planet, heavier matter sank toward the center and lighter material accumulated in the outer layers, all of it kept in constant motion by the intense heat.

And then, when the Earth was a mere 100 million years old, a cataclysmic event changed it forever. In the continuing chaos of the evolving solar system, a large object, perhaps as large as Mars, approached our planet on a collision course. Even a thick, protective atmosphere would not have cushioned the Earth against the devastating impact that followed. The planetoid struck at a low angle, a glancing blow that briefly buried it in the molten mass of Earth's primordial shell. So great was the speed of the object, so huge was the collision, that much of it bounced outward again into space along with a large volume of Earthly matter.

But it did not fly out into space. Slowed, weighed down, and unable to escape the Earth's gravitational field, this giant ring of matter soon coagulated into a single ball that orbited the mother planet. The Earth had acquired its Moon.

Imagine the scene, 4,500 million years ago. Just a few hundred miles above the Earth's surface hung an incandescent Moon that filled the night sky from one horizon to the other, seemingly so close to the Earth that you could touch it. A gaping craterlike depression marked the place where the impact had occurred, threatening for a time the very structure of the planet. The low-angle blow from the impact object set the Earth spinning on a wobbly axis, so fast that one rotation may have lasted only about four hours. The force of this rapid rotation set up wild currents of motion in its outer as well as inner layers.

Yet our planet held together, and the Moon's orbit grew progressively larger during the several hundred million years that followed. By about four billion years ago, the Earth's rotation had slowed significantly as well, so that our planet's day had lengthened to around ten hours, and the Moon was nearly half as far away as it is today. (The Moon continues to move away from the Earth in very small but measurable increments.) At the same time, patches of the Earth's crust began to cool enough to harden molten material into the first solid rocks. Initially, these patches soon were melted down again by streams of hot lava, but eventually some of them survived. The Earth had begun to form a crust.

Not much survives from these ancient "rafts" of solid rock. Earth's continental landmasses are continuously recycled, pushed and pulled below the crust, heated, melted, and regurgitated along midocean ridges and trenches, so it is remarkable that rocks several billion years old do survive in a few places, including Western Australia and interior Africa. But the familiar continental outlines we see on globes and in atlases today are nothing like their antecedents three to four billion years past.

Not only are whole continents recycled: they move horizontally in a process the climatologist-geographer Alfred Wegener called continental drift. A century ago Wegener, observing the close "fit" of the shapes of continents across the Atlantic Ocean, theorized that this was unlikely to be matter of chance. The landmasses had once formed part of a supercontinent, he reasoned, that fractured into the continents we see on the map today. He called this hypothetical supercontinent Pangaea, and his map of (Fig. 3-1) is one of the most prescient ever drawn (Wegener, 1915).

Wegener's theory engendered the later theory of plate tectonics and crustal (sea-floor) spreading. Scientists now know that Pangaea and its breakup were

CONTINENTAL DRIFT

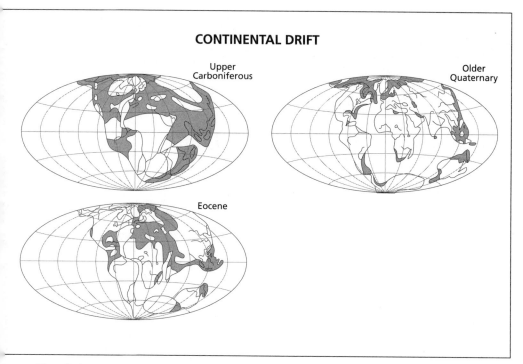

Fig. 3-1

only the latest episodes in a cycle of continental coalescence and splintering that spans billions of years. This latest Pangean fragmentation, however, began only about 180 million years ago and continues to this day. Where great slabs of the crust called plates collide and continental margins are pulled under, as is happening along the west coast of North America, earthquakes and volcanic eruptions accompany the process. That is why we recognize a circum-Pacific "Ring of Fire," marking these gigantic collisions from Chile to Alaska to Indonesia to New Zealand (Fig. 3-2). The earthquake that caused the December 26, 2004, tsunami (its epicenter is marked by the arrow off northwest Sumatra) was the result of such plate collision.

So the continents, made of the lightest rocks (solid or molten) on Earth, ride like rafts on the mobile, heavier plates below. And what makes those slablike plates move? That was Wegener's unsolved problem: the mechanism for his continental-drift theory. The answer came from an unlikely place: the ocean floors, where upwelling, red-hot lava creates new crust even as old crust elsewhere is pushed under (the process is called subduction) as plates collide. Wegener kept comparing North America and Europe and South America and Africa, but the crucial answer to his unsolved problem lay midway between them: along the Midatlantic Ridge, where the Transatlantic

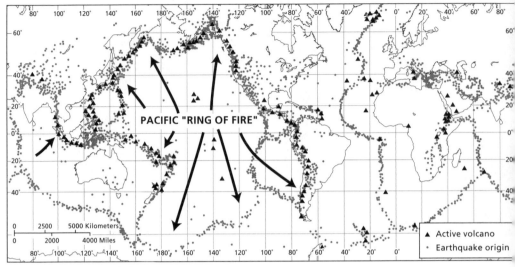

PACIFIC "RING OF FIRE"

▲ Active volcano
• Earthquake origin

Fig. 3-2

continents were once conjoined. There, heavy, dark, basaltic rock forces its way upward, pushing earlier, hardened lava aside. We can actually observe the process occurring today near Iceland, where the normally submerged Midatlantic Ridge rises to the surface. New land is being formed as islands of lava rise above sealevel.

All this is not just theory any longer. We can now measure the movement of continents. One year from now, the room in which you are reading this will be about a half inch (13 mm) from where it is today, assuming you are somewhere in North America. That may not seem to be much, but calculate what it means over geologic time. In just 1 million years, the distance will be 8 miles (13 km). But the breakup of Pangaea started about 180 million years ago—and the North American plate is by no means the fastest-moving one. So the continental landmasses, and the plates that carry them, have moved thousands of kilometers since then.

What will the map of the distant future look like? At present the Earth exhibits the kind of pattern that attracts geographers' attention: so uneven is the distribution of landmasses that we routinely speak of a "Land Hemisphere" and a "Sea Hemisphere." This, of course, was far more pronounced when Pangaea still existed. Plate motion since then has already lessened the Land-Sea Hemisphere contrast. But the Pacific and its oceanic plates still cover nearly half the planet's surface, and the process has a long way to go. It may be, however, that plate movement is slowing down, and some geologists and others working on this problem suggest that continental motion may actually come to a halt, after which the landmasses may once again converge to form still another supercontinent. So "continental drift" may in fact be a cyclic process that has been moving the landmasses for as long as the Earth's crust has had plates and continents.

There is some evidence for this notion here in North America. As we noted, when plates converge and collide, the lighter continental plate over-rides the heavier oceanic plate, but parts of both are pushed downward during subduction, and continental crust, with its fossils and telltale structures, is lost. That is happening now along North America's west coast, where spectacular coastal scenery bears witness to the forces at work. But before Pangaea broke up and North America started its westward journey, the continent was headed the other way, being carried toward the Pangaean cluster of landmasses. The coast of today's eastern United States was jammed into Morocco and other parts of West Africa, and the East, not the West, had the scenery associated with subduction. The Appalachian Mountains bear eroded witness to that pre-Pangaean time.

Over the long term, therefore, think of our planet's surface as ever changing, of continents moving and the crust shaking, of oceans and seas opening and closing, of land lost by subduction and gained by eruption. And this is only one dimension of the ceaseless transformation of Earth that began 4.6 billion years ago.

OCEANS PAST AND FUTURE

The fall of the Berlin Wall in 1989 led to much introspection—not only political, but also philosophical and scientific—and gave rise to a spate of books signaling the onset of a new era. Their titles were often misleading, such as *The End of History* by Francis Fukuyama, but none more so than one by John Horgan (1996) called *The End of Science*, which argued that all the great questions of science had been answered and that what remained, essentially, was a filling of the gaps. When it comes to global environments, however, some great questions remain open.

One of these relates to the oceans. Planet Earth today is often called the Blue Planet because more than 70 percent of its surface is covered by water and views from space are dominated by blue hues and swirls of white cloud, but in truth we do not know with any certainty how the Earth acquired its watery cloak, or exactly when. Some scientists hypothesize that the water was originally trapped inside the Earth during its formation and rose to the surface during the time when heavier constituents sank to form the core. The gases that are released during volcanic eruptions are mostly (more than 95 percent) water vapor, and massive volcanism marked Earth's early history, though lessening over time. Others calculate that most of the water that did reach the surface in this way would have been evaporated into space by the searing heat then prevailing, suggesting that another source must be identified. This has led to the comet hypothesis, which proposes that icy comets bombarded the Earth for more than a billion years while its atmosphere was still thin, accumulating fresh water from space that filled the basins in the formative crust. But studies published in late 2004 report that the chemistry of the oceans cannot be matched to that of icy comets (now much better known than before), casting doubt on the comet hypothesis.

Obviously, the "end of science" has not arrived when it comes to as crucial a question as this, and here is a related one: will the Earth retain its life-giving oceans permanently? Probes of our neighboring planet Mars produced some startling conclusions: Mars may have lost a global ocean more than 30 meters (100 feet) deep, and there are indications that Mars at one time had even more water (as a proportion of mass) than planet Earth. Why and how rapidly did Mars lose its ocean? And what may that loss portend for Earth?

ICE ON THE GLOBE

I always suggest to my students that they should be as familiar with the geologic time scale as they are with the months of the year and the days of the week—it is a great way to keep things geographic in temporal perspective. (Table 3-1). Geologists refer to the first 800 million years as the Hadean, and indeed the Earth was hot as Hades during that eon. The next 1,300 million years form the Archean, when the oldest surviving continental rocks and the first life forms are recorded. Next comes the Proterozoic eon, lasting from 2,500 million until 570 million years ago. It is late during this eon that something dramatic appears to have happened: the Earth went into a deep freeze.

The theory is known as Snowball Earth, and the evidence is coming from rocks as far apart as China and Australia. It suggests that the Earth did not just cool, as has happened several times since: apparently the entire planet froze, from pole to pole and from land to sea. The landmasses lay buried under ice and snow; the ocean surface was frozen solid. What might have caused this to happen? A temporary but significant decline in the Sun's radiative output is one possibility. A rapid decline in methane-producing microorganisms (methane was the key early greenhouse gas) with the rise of oxygen-generating microbes around 2,300 million years ago may have chilled the entire planet (Kasting, 2004). Some scientists suggest that the stabilizing crust and consequent decline of volcanic activity could be a factor. Whatever the cause,

THE GEOLOGIC TIME SCALE

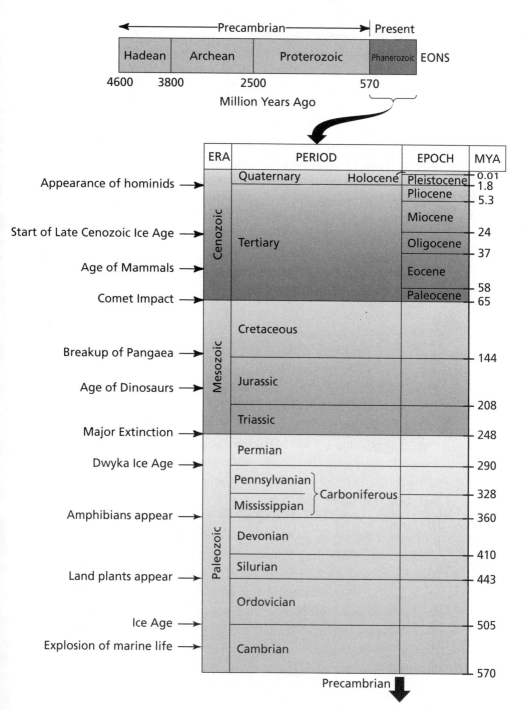

Table 3-1

the planet and its early life forms experienced a crisis. Whether the Earth was indeed a "snowball" or, as others suggest, it was a less frigid "slush ball," the days of sustained warmth as its hallmark were clearly over.

Whatever the outcome of the search for evidence relating to the Snowball Earth theory, we know that this was not the last time the Earth experienced an ice age. Several have followed, and the most recent one is in progress right now. All known ice ages, and perhaps even the Proterozoic one, have periods of severe cold separated by shorter phases of comparative warmth. We are experiencing such a warm interruption at this time, one that has lasted roughly 12,000 years—roughly, because the rapid warming that brought us today's mild climates actually started about 18,000 years ago but was sharply interrupted about 12,000 years ago, when frigid temperatures briefly returned. But the ice age of our time began about 40 million years ago and, as we will see later, continues.

What we do know about ice ages is that their often rapid environmental swings pose challenges to all life forms, from eukaryotes to humans. Ice ages are times of accelerated evolution; organisms that adapt tend to survive, those that cannot, perish. During the Snowball Earth Ice Age, single-celled eukaryotes evolved into multicelled, more complex organisms. Those with protective shells (there was plenty of calcium carbonate around) had a better chance of survival. When the ice age ended, life on Earth was set for its Cambrian Explosion, the burgeoning of marine organisms in unprecedented diversity that marks the oldest period of the Paleozoic era. Then, after more than 4,000 million years of the planet's existence, began the drama that would lead to the emergence of humanity, 570 million years later, again during an ice age.

How many ice ages have occurred between the Proterozoic and the current one? We are sure of one at least: the ice age that occurred when Pangaea as a supercontinent was still in one piece (Fig. 3-3), and it may have been unprecedented in its impact on life on Earth. This happened during the Permian period, the last of the Paleozoic era, between 290 and 251 million years ago. At the time, Pangaea consisted of a northern periphery called Laurasia, consisting of parts of Eurasia and North America, and a southern sector known as Gondwana, at whose core lay what is today Africa. The South Pole was just offshore from South Africa; Australia adjoined India and Antarctica. When the Permian Ice Age (known elsewhere as the Dwyka Ice Age) struck, forests were widespread, amphibians thrived, small reptiles had made their appearance, and insects had proliferated. When it was over, one of the greatest mass extinctions ever had decimated life on Earth. As the map shows, a vast area of Gondwana was buried under ice, but what the map cannot show is the ice age's impact on environments far beyond the

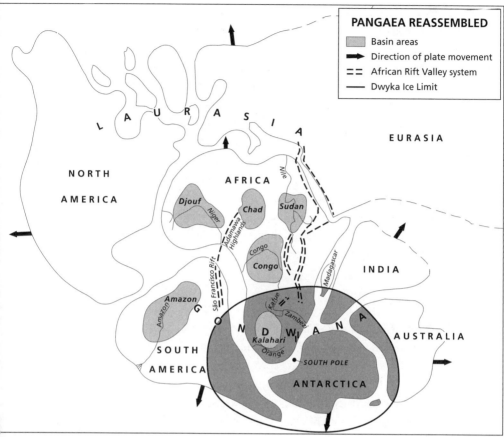

Fig. 3-3

high latitudes. Huge areas fell dry, forests withered, and countless species of plants and animals became extinct.

That, at least, was the state of knowledge until November 2003, when geologists produced evidence indicating that the Permian period ended in catastrophe, not frigidity. Asish Basu and a team of researchers found tell-tale bits of meteorite in Antarctica along with other indications that a huge meteor struck the Earth about 251 million years ago, killing as much as 90 percent of all life on the planet (Kerr, 2003). Rather than succumbing to ice-age cold alone, animals as well as plants were incinerated by the impact and its aftermath, which may have included huge volcanic eruptions pouring out vast volumes of molten rock from fissures and vents in the shaken crust. The scientists theorize that this may have been the most devastating of the Earth's five known mass extinctions, resulting from a combination of environmental, extraterrestrial, and geological forces that ended not only the Permian period but also the Paleozoic era (Table 3-1).

When the ice age ended and the Mesozoic era opened, there was little left of Permian life. But now the post-ice-age planet made up for that. Tropical warmth replaced Arctic cold, moisture and precipitation abounded, atmospheric oxygen increased markedly as luxuriant forests spread and the Earth was ready for the faunal exuberance of Jurassic Park. The age of the dinosaurs also saw the first birds, the first marsupials (animals whose females nurture their offspring externally in a pouch, not internally in a placenta), and the first angiosperms (plants whose seeds are encased in fruit).

Even the massive, sky-blackening volcanism that accompanied the breakup of Pangaea during the Jurassic failed to spoil the party. The dinosaurs grew larger and larger, specializing into herbivores and carnivores and competing fiercely for survival. As the landmasses separated and the seas between them widened, species found themselves isolated and evolved into distinctive forms. Only another ice age, it seemed, could end the Mesozoic's profusion.

SUDDEN DEATH

The diversity of dinosaurs and the flourishing of Mesozoic plants reached their zenith during the Cretaceous period, when huge birds flew in vast forests and flowering plants spread across the world. It was warmer even than it is today; polar and mountaintop ice were long gone and dinosaur species roamed from Alaska to Antarctica. Small mammals managed to survive in special niches, but the day belonged to the giant reptiles whose only enemy, it seemed, would be a sudden return to the frigid conditions of the Permian.

But the age of the dinosaurs came to a much more dramatic end—not with a glacial whimper but with an extraterrestrial bang. One day about 65 million years ago, a comet or asteroid only about 6 miles (10 km) in diameter streaked toward Earth at a speed of 55,000 mph (90,000 kph) on a collision course. It approached from the southeast at a low angle, striking Earth in what is today the area of the Yucatan Peninsula of Mexico. When you get off a boat at the small port of Progreso, there is a small, hand-painted sign that points to Chicxulub, a Maya name for a local village. But to geographers, Chicxulub means the end of one era and the start of another. Here the asteroid's impact produced an explosion equivalent to about 100 trillion tons of dynamite, forming a crater approximately 110 miles (180 km) in diameter, 40 miles (65 km) deep, and encircled by a geological fault 20 miles (about 30 km) beyond, all of it buried today by later sediments.

It is possible that the Chicxulub asteroid was one of a swarm, and that smaller ones struck the Earth elsewhere, including the ocean. In any case, the impact's devastation reached around the planet, and was at its worst in North America. The impact area was a shallow sea with soft, deep sediments, and the blast sent a mass of debris hurtling thousands of miles into the heart of the continent and high into the atmosphere and beyond. Researchers David King and Daniel Durda calculate that some of it reached halfway to the Moon before falling back to Earth. And when it did fall back, it rained red-hot rocks on the rotating planet, setting fires to forests almost everywhere. The atmosphere was heated enough to evaporate entire lakes, incinerate whole ecosystems, and extinguish most life over large low-latitude regions.

The Chicxulub impact ended the Cretaceous and marked the beginning of a new geologic-calendar period, the Tertiary. Popularly, the transition is called the K/T Boundary, but its significance is hard to overstate, because this was one of the three greatest known mass extinctions ever. While it is possible that some dinosaurs survived the original blast, notably in higher latitudes, food chains had been fatally disrupted and they, too, died out. Some small mammals were better equipped to outlive the crisis, perhaps keeping cool in high-latitude caves and burrows, depending less on the luxuriant vegetation and reptilian life the dinosaurs had needed. But the faunal and floral exuberance of the Mesozoic era came to a sudden, irrevocable end.

The K/T blast had long-term effects on global environments. Much of the enormous volume of pulverized, ejected rock remained in orbit around the Earth, choking the atmosphere and blocking the sun. The smoke from worldwide fires darkened the skies worldwide. Eventually the overheated atmosphere cooled, and the blockage of the sun sent temperatures plummeting still more, creating colder global conditions than had been experienced for 185 million years—since the Permian Ice Age. Now it becomes important to be familiar with the epochs of the Tertiary period, because the first of these epochs, the Paleocene, witnessed major climactic reversals, and the next one, the Eocene, saw the beginnings of a new ice age that probably would have come whether the K/T impact occurred or not.

BACK TO THE FUTURE

As the post-impact planet cooled, shrouded in dust and smoke, there was little to suggest that an era of recovery and renewed biodiversity lay ahead. The forest fires, and the explosion into the atmosphere of huge volumes of carbonates from the impact site, greatly raised the amount of carbon dioxide in the air, creating a powerful greenhouse effect when the skies began to clear. As Figure 3-4 shows, this global warming continued through much of the Paleocene, raising temperatures even higher than they had been during the warm Cretaceous. Biogeographers conclude that this killed many plant and animal species that might have survived the blast and its immediate aftermath. But the Paleocene's warming did not continue. The next epoch, the Eocene, was marked by an almost continuous drop in global temperatures, and by the time it ended, 36 million years ago, permanent ice was beginning to form on the Antarctic continent. The Cenozoic Ice Age was about to start.

Soon the evidence began to accumulate: the early phase of the Oligocene witnessed the beginning of the formation of the Antarctic Ice Sheet even as South America and Antarctica were separating. (Remember: through all of this activity, the continents continued to move on their crustal plates, the Atlantic Oceans, North and South, kept widening, and the distribution of land and water on the planet kept changing.) Even before the ice on Antarctica reached its shores, glaciers began to develop on the Earth's highest mountains, filling high-elevation valleys and sculpting a new, angular topography of sharp-edged peaks and ridges. Tree lines dropped to lower altitudes, vegetation shifted equatorward, and the mammals that had become the dominant species, including the now-common primate forms, responded by migrating and adapting as environments fluctuated.

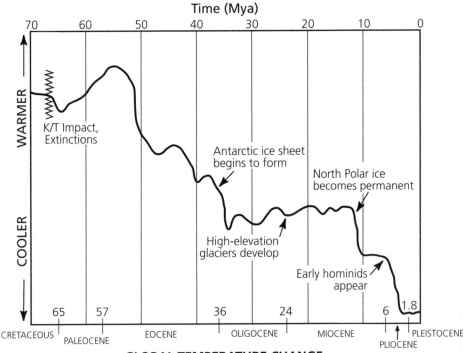

GLOBAL TEMPERATURE CHANGE,
LATE CRETACEOUS TO PRESENT

Fig. 3-4

Ice ages are not uniform cooling events: surges of coldness and advances of glaciers are interrupted by temporary warming spells long enough to reverse much of the glacial impact. So it was from about the middle of the Oligocene to the middle of the Miocene when, it seemed, the Cenozoic Ice Age had reached an equilibrium that (had weather and climate analysts been around in those days) might have been taken as a sign that the worst was over. Antarctica still had coastal zones clear of ice; the high-mountain glaciers advanced and receded as global climate cooled and warmed, and while the planet overall was cooler and drier than it had been during the ages of the dinosaurs, there was plenty of environmental variety and related biodiversity. But then, about 14 million years ago, global cooling resumed with a vengeance. Antarctica's ice sheet not only reached the ocean all around its shores, but the ice floated from land onto water and cooled the Southern Ocean, affecting the entire global ocean in the process. Permanent ice appeared and rapidly thickened on the waters of the North Pole and environs. The temperature plunge continued until a comparatively brief period of stabilization marked the end of the Miocene, but the next, brief (4.2-million-year) Pliocene epoch saw another period of further cooling. Permanent glaciers appeared even on mountains in equatorial zones of the Andes, East Africa, and New Guinea. About 1.8 million years ago began what was on average the coldest epoch of the Cenozoic Ice Age, the Pleistocene. We are living under Pleistocene conditions today, enjoying the autumn of a warm phase that, as noted earlier, has been going on for about 12,000 years.

CLIMATES AND PRIMATES

An ice age is a long-term event, lasting tens of millions of years and bringing profound changes to all parts of the planet, not just those directly affected by advancing ice sheet and valley-filling glaciers. The overall process and its manifestations may operate slowly, but there are times when sudden surges of advancing ice move fast enough to encircle grazing animals and snap off mature trees like matchsticks.

Advances of ice into lower latitudes and altitudes during a cold ice-age period are called glaciations and temporary warm-ups between glacial advances are referred to as interglacials. It should not be necessary to emphasize these distinctions, but geologists and even some physical geographers occasionally get this wrong. One of the leading geology textbooks, for example, states that "periods during which the average temperature at the Earth's surface dropped by several degrees and stayed low long enough for existing ice sheets to grow larger . . . are called glaciations (or ice ages . . .)." What should have been said, of course, is that *during* an ice age, such temperature declines are marked by glaciations (Murck & Skinner, 1999). This is important, because even while global temperatures decline during an ice age and glaciations push ice into ever-larger areas, there are also periods of relief in the form of interglacial warmth and retreating ice (a better word is *receding*, because glaciers do not really reverse direction).

Let us keep in mind, therefore, that the overall drop in temperatures that began in earnest during the Eocene and reached unprecedented lows during the most recent, Pleistocene glaciations, was no steady or continuous process. As Figure 3-4 suggests, there were long periods when the Cenozoic Ice Age seemed to have reached its maximum, with its end in sight—only to plunge into even more frigid conditions. What is certain is that the impact of that ice age was evident all over the world. When a vigorous glaciation pushed ice sheets deep into present-day Canada and into lower latitudes of Eurasia, it got cooler and drier even in Africa and equatorial South America. Tropical forests shrank, savannas expanded, and animals as well as plants responded by migrating and adapting to new environments.

We all know that the story of the great apes, hominids, and humans played itself out in Africa. As another chapter in this book reminds us, all of us are, ancestrally, Africans. But any geographer looking at the world map would wonder: why Africa, when Eurasia is so much larger and seems to contain so much more environmental diversity? In that connection, one question that troubled me for decades after I first learned of it in graduate school a half century ago: why are chimpanzees and orangutans, two of humans' closest genetic relatives, separated by thousands of miles of land and ocean, the chimps in Africa and the orangs in Southeast Asia?

I remember sitting in an airport van full of archeologists and anthropologists going home from a meeting in Asheville, NC about five years ago, and raising this issue. Where, I asked, is the fossil record that would prove the migration of orangutans from Africa to Southeast Asia? It will be found, I was told. Fossils in humid South Asia don't survive like they do in Africa. We're talking 6, maybe 7 million years ago. The genetic relationship is beyond doubt, so the migration must have taken place. It's just a matter of time before the evidence appears.

Well, the evidence never surfaced, and I should have thought more carefully about the geographic implications of the question I raised. If the great apes of Southeast Asia did not descend from those of Africa, then each branch mush have had ancestors in Eurasia. Postulate that, and you conclude that something drove one of those branches—the one leading to the gorilla and chimpanzee among others—to tropical Africa, while the other, leading to the orangutan, was pushed into tropical Southeast Asia.

What could have been the impetus for this dual migration? The worsening of Cenozoic Ice Age conditions, of course. Before Miocene climatic conditions deteriorated, the global environment, though cooler than it had been during the Oligocene, remained fairly stable (Fig. 3-4). During that time, the descendants of *Proconsul* and other early apes probably moved out of Africa into environmentally diverse Eurasia, where forest habitats varied widely and relatively warm temperatures ensured an ample supply of fruits and other forage. Now Eurasia, not Africa, was the heartland of differentiation and adaptation for the great apes, and numerous lineages evolved, some of them now part of the known fossil record.

But in the late Miocene the Cenozoic Ice Age took a turn for the worse, dropping global temperatures, freezing over the Arctic Ocean, drying up vast stretches of once-forested Eurasia, and destroying habitats that had for millions of years nurtured the great-ape families. Extinction was commonplace, but researchers have determined that two lineages managed to survive by adaptation and migration: *Dryopithecus*, which occupied southwestern Europe, and *Sivapithecus*, which was based in the forests of the northern Ganges River basin (Fig. 3-5).

Dryopithecus moved southward across what is today the Mediterranean into tropical and eventually equatorial Africa, probably around 9 million years ago, adapted to cope with the severe swings of environment prevailing at the time and destined to give rise to tool-making, large-brained descendants. It is somewhere along the *Dryopithecus* lineage that hominids and Africa's great apes had a common ancestor, but make no mistake: having escaped the rigors of the late Miocene in Eurasia, they found no African Garden of Eden. The increasing severity of the Cenozoic Ice Age, persisting into the Pliocene and the Pleistocene, affected tropical and equatorial Africa as well, causing rapid environmental swings that overwhelmed and extinguished numerous progenies, ape and hominid alike (Begun, 2003).

The other Eurasian great-ape lineage, *Sivapithecus*, moved down the Malayan Peninsula and into what is today Indonesia, where Miocene and Pliocene conditions may have been less rigorous than they were in ice-age Africa. In any case, no comparable evolutionary drama occurred here: the orangutan does not share a hominid ancestry and is the end of its line. Eventually and ironically, descendants of *Dryopithecus* and *Sivapithecus* would come face to face—but not ape to ape. When interglacials in the late Pliocene warmed the Earth enough to revive forests and refill desiccated lakes, Africa's hominids did what early Miocene apes had done before them: migrate

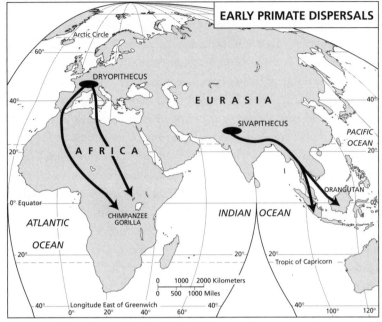

Fig. 3-5

out of Africa into Eurasia. And so the first of these emigrants, *Homo erectus*, moved across Arabia and southern Asia and, one day somewhere in Malaya or on Java or Borneo, saw a creature that, if the trip had been swifter, would have reminded her of the chimpanzees and gorillas left behind in Africa. Orangutan and hominid had closed a 9-million-year circle.

THE FRIGID PLEISTOCENE

Earlier we noted that the temperature plunge that began in the late Miocene and continued during the Pliocene (when early hominids made their appearance in Africa) set the stage for the Pleistocene epoch, beginning less than 2 million years ago with a series of the most severe glaciations of the entire Cenozoic Ice Age interrupted by short, warm interglacials. In Africa, where hominid evolution was under way, these phases were marked by fluctuating climates and sometimes wild fluctuations in ecologies that guided natural selection. *Homo erectus*, successor to *Australopithecus*, was the most successful of these hominids. *Homo erectus* managed to cope with forests that changed into savannas and back again, lakes that formed and evaporated, wildlife that ranged from easy to difficult and elusive prey and, as the fossil record shows, even with massive volcanic eruptions and huge ashfalls. Undoubtedly the numbers and dispersal of *H. erectus* also varied, but the species survived for perhaps as long as 2 million years and, as we noted, left Africa and spread across Eurasia, reaching not only Southeast Asia but also present-day China in the east and Europe in the west. *Homo erectus* and its successor, the tool-making *Homo habilis* ("handy man"), presaged the human expansion that was to follow similar paths.

Pleistocene environmental conditions are the subject of intense investigation today. When Pleistocene glaciations were severe, permanent ice advanced deep into the landmasses of the Northern Hemisphere (Fig. 3-6). This drastically changed the distribution of plants and animals, shifting them equatorward latitudinally and downslope altitudinally. Faunal ranges and refuges shrank, niches became unusable, and always there were species that failed to survive the transition. Such glaciations could last as long as a hundred thousand years, but eventually an interglacial would warm the planet, melt much of the ice, and living space as well as survival opportunity would expand again.

Physical evidence from various sources, including Greenland ice cores and Atlantic ocean-bottom mud deposits, coupled with analyses of broken and pulverized rocks left behind by the Pleistocene glaciers, at first seemed to suggest a remarkable regularity in the ups and downs of Pleistocene temperatures. Interglacials seemed to last an average of 10,000 years, so that, on average, a glaciation-interglacial sequence encompassed around 110,000 years or so. Over the past 425,000 years of the Pleistocene, there appeared to have been four major glaciations followed by four interglacials, the latter including the current one, the Holocene.

More recent analyses suggest that it has not been so simple. The most recent glaciation, the Wisconsinan, began about 100,000 years ago after a rather long interglacial, the Eemian. But the Wisconsinan was not one long cold spell. In fact, it was punctuated by several brief interglacials and longer (comparatively) mild spells that made habitation in higher latitudes possible for thousands of years. What is clear is that these alternations from warm to frigid and from mild to cool often happened quite suddenly, taking their toll not only on animals and plants but also on hominids and humans.

We now begin to put humans in the picture, because they appeared in Africa some time during the glaciation preceding the Wisconsinan, probably around 170,000 years ago. Hominids and humans shared not only parts of Africa but also the migration routes from Africa into Eurasia, and they met similar fates when the climate took its disastrous turn. We know something about this because early humans used the land bridge between Africa and Eurasia, across the Sinai Peninsula, late during the Eemian interglacial (Fig. 3-7). No

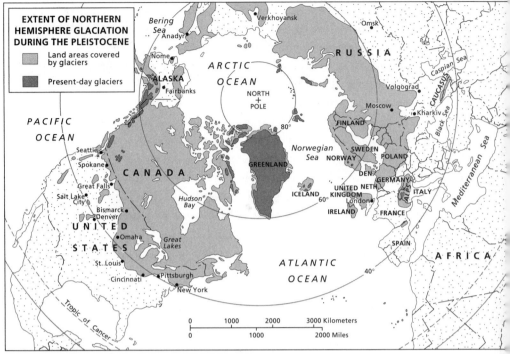

EXTENT OF NORTHERN HEMISPHERE GLACIATION DURING THE PLEISTOCENE

Land areas covered by glaciers

Present-day glaciers

Fig. 3-6

sooner were they in the Middle East when the Wisconsinan Glaciation began with a ferocious drop in temperatures. All that is left of that early emigration are the bones of the migrants. They never made it to Europe.

The next time humans tried to leave Africa they used a different exit at the opposite end of the Red Sea around 85,000 years ago. The Wisconsinan Glaciation had converted so much water into ice that the surface of the Red Sea was hundreds of feet lower than it is today. The reefs in the Bab-al-Mandab (Arabic for "Gate of Grief"), where the Red Sea opens into the Indian Ocean, created the stepping stones that facilitated the crossing, and our African ancestors were on their way. First they moved along the shore of the Arabian Peninsula, then around the Persian Gulf and on into India and Southeast Asia, reaching Australia via New Guinea (another crossing facilitated by low sealevels) about 60,000 years ago (Oppenheimer, 2003).

In the process, modern humans met the hominids who had preceded them into Eurasia, and the hominids were no match for the resourceful newcomers. When the first modern humans (the Cro-Magnons, as they are known) reached Europe from India via the Middle East and found the Neanderthals, and earlier *Homo* species, on the scene, they quickly overwhelmed them with their complex culture ranging from cave art to tool kits and from

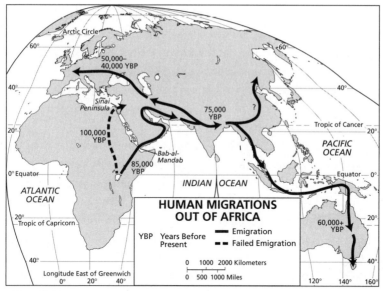

Fig. 3-7

inventive fishing gear to sewn clothing. They lived in cooperative communities and used sophisticated language, and thus modern humans had all the advantages over the survivors of earlier out-of-Africa migrations. Their technology gave them the opportunity to cope with the Wisconsinan's climatic swings: in milder times they expanded their frontiers, and when it got colder they devised ways to cope with increasingly harsh environments. That is why paleoanthropologists are finding fossil evidence of human settlements in Europe that survived through pretty severe cold periods. Humans were finding ways to combat the rigors of changeable climate.

A CLOSE CALL

In truth, humanity was fortunate to be able to expand into Europe at all, because the exodus from Africa via the Arabian Peninsula and South Asia almost came to a halt in spectacular fashion. We are not sure just where the vanguard of human migration was about 73,500 years ago, but the exodus across the Bab-al-Mandab was probably continuing. Then a catastrophic event in what is today Indonesia very nearly wiped all of humanity off the planet. On the island of Sumatera (Sumatra in the old spelling), a volcanic mountain now named Toba did not just erupt—it exploded. This explosion sent millions of tons of debris into orbit, obscuring the Sun, plunging much of the Earth into long-term darkness and altering global climate. Mount Toba's detonation could not have come at a worse time. The Wisconsinan

Glaciation was in full force, the Earth's habitable zone was already constricted, and a large part of the still-sparse human population faced death. Anthropologists refer to this event as humanity's "evolutionary bottleneck," suggesting that a great deal of genetic diversity must have been lost in that instant. Today, the filled-in-caldera marking Toba's cataclysm is 55 miles (90 km) long and 30 miles (50 km) wide, silent witness to the greatest threat to human survival since we emerged on the African savanna.

As has happened so many times in Earth's history, the skies eventually cleared, the atmosphere was cleansed, and normal conditions resumed. Toba was calamitous, but it was no Chicxulub. It did not generate fires around the world, and was not nearly as destructive as that incoming asteroid. Nevertheless, it posed a real danger to humanity, and reminds us that risks of this kind have not disappeared. Geologists refer to the Toba explosion as a 500,000-year event, something that happens, on average, once in a half-million years. That is no guarantee that another Toba-like eruption will not happen for a very long time. Our planet still poses unpredictable, incalculable natural hazards, as we were reminded on December 26, 2004.

So let us anticipate the geographic story of climate and civilization by revisiting the environmental saga of the Cenozoic Ice Age. For tens of millions of years, glaciers have been spreading during cold periods appropriately called glaciations, only to recede during warm spells we know as interglaciations. The entire planet and all its life forms, not just the polar regions, are affected by these environmental swings. Glacial advances pushed primates from cooling Eurasia into Africa and Southeast Asia; interglacial warm spells allowed hominids to leave Africa and survive in temperate Eurasian latitudes. All the while, average global temperatures declined until, about 1.8 million years ago, the start of the Pleistocene brought an alternation of long glaciations and short interglaciations. About 120,000 years ago, the Eemian interglacial was even warmer than the current one, the Holocene, but it ended abruptly some 110,000 years ago with a rapid advance of the ice of the Wisconsinan Glaciation. Humanity was on the scene by then, but our earliest emigrations from Africa were stymied by the sudden return to frigid conditions. When we finally made it out of Africa, we kept a southern route along Asian shores, but there were times when the Wisconsinan cold ameliorated and our ancestors took these opportunities to penetrate present-day Europe and confront—and overpower—Neanderthal predecessors. But the relentless advances of the ice came again, and just 20,000 years ago, glaciers stood as far south as the Ohio River in North America and southern England, central Germany, Slovakia, and Ukraine in Europe. Then, about 18,000 years ago, global warming sent those glaciers into fast recession, so fast that whole regions rapidly emerged from under the ice, huge ice sheets slid into the oceans, the sealevel rose, the margins of the continents were submerged, land bridges between continents and islands were inundated, and the map of the physical world began to look similar to the one we know today. Twelve thousand years ago, cold conditions made a brief comeback, but that did not last. From about 10,000 years ago until today, humanity has thrived in the warmth of a prolonged interglacial we call the Holocene, but unlike the Eemian, the Holocene has witnessed the emergence of complex cultures and civilizations, the population explosion, the formation of states and empires, the growth of megacities, and the burgeoning of technology in countless forms. It has also seen wars and destruction on an unprecedented scale. With our human numbers approaching 7 billion and global warming opening the last niches for habitation, the question is: what happens when the ice returns, as it has more than two dozen times during the Pleistocene?

4

CLIMATE AND CIVILIZATION

When the long dominance of the Wisconsinan Glaciation finally came to an end and the vast, thick glaciers that had covered all of Canada and most of the United States Midwest began to recede, planet Earth embarked on an environmental transformation not seen for more than a hundred thousand years. Global warming, starting about 18,000 years ago, was more than a temporary respite of the kind human communities living in high latitudes had experienced before. This time the warming was so powerful and persistent that the glaciers soon gave way, not just along their leading edges but all over, melting fast and yielding enormous volumes of water that raised sealevels rapidly. Streams emanating from the ice became roaring rivers carrying a huge load of sediment (the Mississippi Valley and Delta are filled with it). While sealevel was still low, those rivers, like the Hudson, carved canyons in the continental shelf, soon to be drowned by the rising water but evidence of the physiographic turmoil that attended the great melting. In Europe, Britain was severed from the mainland by the English Channel; meltwater poured from the glaciers on the Scandinavian Peninsula and filled the Baltic. Ice disappeared from central Germany and even most of Russia. The glaciers in the Alps receded as the ice cap over the mountains thinned under the hot sun.

To our Stone Age ancestors who had been living with the Wisconsinan's variable climate, this warming must have been a welcome experience. They had been in Europe for tens of thousands of years, shifting their abodes as the environment fluctuated (some scholars hold that Native Americans arrived in the Americas much earlier than about 13,000 years ago, still the majority view). But this time the warming was so persistent that people ventured farther and farther poleward. Warmer summers and milder winters opened opportunities for more secure livelihoods and larger communities.

ONE FINAL SURGE

But nature still had a surprise in store. By about 12,000 years ago, the ice had melted from southern regions but still held on in parts of Canada and northern Europe. Because these ice sheets were lubricated, even at their bases, with meltwater, the inevitable happened: they started to slide downslope, that is, coastward and into the ocean. One especially large ice sheet, the size of a large Canadian province, slid into the North Atlantic, causing disastrous waves along coastlines from Europe to the Caribbean and chilling the ocean right back to glaciation-time temperatures. This event, called the Younger Dryas after a Tundra wildflower that flourishes where the ice once stood, caused more than a thousand years of cooling (our ancestors must have said "here we go again"), but this time that cooling was temporary and brief. By about 10,000 years ago, temperatures were back on the global-warming track, the remaining ice was melting again, and the Younger Dryas was just a hiccup in the overall pattern.

HOLOCENE HUMANITY

The warm period that has now lasted more than 10,000 years is referred to as the Holocene Epoch by geologists, but of course there is no evidence as yet that it is in fact an epoch, that is, a new phase following the end of the Cenozoic Ice Age. All the Holocene is, as far as can be known, is another Pleistocene interglacial. What is different about it is not geologic, but geographic. About the fact that this is a *cultural* epoch like no other there can be no doubt: hominids have been around for millions of years and humans for maybe 170,000, but nothing like the demographic or cultural explosion witnessed by the Holocene has ever happened before.

Before the Holocene took on its present environmental character and its climate to some degree stabilized (it never stopped changing, but within narrower parameters), still another drama unfolded. Geologists report that a final mass of ice sliding off northernmost North America into the Atlantic not only cooled the ocean again, though far less than the Younger Dryas event, but also sent a wall of water coursing through the Strait of Gibraltar into the Mediterranean Sea. The fast-rising Mediterranean then overflowed its barrier with the Black Sea, on whose shore stood numerous villages. Geologists William Ryan and Walter Pitman describe this event as having the force of 200 Niagara Falls, filling the Black Sea at the rate of 6 inches (15 cm) per day and forcing the coastal-plain inhabitants living near the shore to back away about 1 mile daily. Many sought refuge in the mountains along the (present-day) Turkish coast, but thousands of others must have abandoned their dwellings, boats, and fields and watched the flood swallow them up. By the time it was over, the surface of the Black Sea had risen 500 feet (145 m) and, quite possibly, the biblical legend of the Great Flood was born.

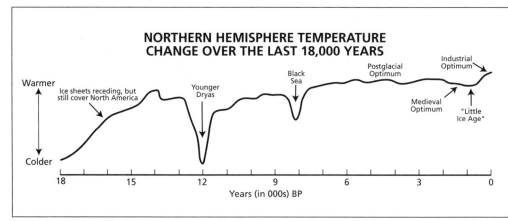

Fig. 4-1

The Holocene, therefore, was no uneventful transition to warmer and calmer times. Still, conditions did stabilize enough following the Black Sea crisis to allow climatologists to talk of a Postglacial Optimum, starting about 7,000 years ago and marking the time when global environmental conditions became rather like those with which we are familiar today (Fig. 4-1). This Postglacial Optimum transmuted into another salubrious period, the Medieval Optimum, starting around 2,000 years ago, which witnessed the expansion of settlement in northern latitudes of Eurasia, the peopling of Iceland, even the colonization of Greenland.

Not all was quiet during these climatically optimal times. About 3620 BP (Before the Present) the volcanic island of Thira (Santorini), to the north of Crete in the Greek archipelago now known as the Cyclades, blew up in a Toba-like eruption that covered a wide area of the eastern Mediterranean with a thick layer of poisonous ash. Days of darkened skies and seismic waves in the waters may have given rise to biblical allusions to darkness and parting waves (and perhaps to the legend of Atlantis, for when daylight returned, Thira was mostly gone). But more consequential is what happened on Crete, where the powerful and culturally advanced Minoan civilization was based. Thira may have dealt it a fatal blow, opening the way to the regional dominance of ancient Greece.

Nor was this optimal period a time of invariably favorable climate. Deglaciation (the recession of ice and the opening of land to atmospheric conditions) continues long after the ice has disappeared: the poleward shift of climatic zones, the maturing of soils, the migration of plants and animals keep altering the environment for thousands of years. Some societies found themselves in favorable locales and converted their good fortune into security, expansion, and power. Others, including some early states and cities in what is today the Middle East, saw rivers dry up, deserts encroach, and livelihoods destroyed. Was the innovation of irrigated farming a response to these challenges, as some scholars suggest, or had population pressure in stable environments led to it previously?

The Medieval Optimum certainly was a good time for two contemporaneous empires: the Roman in western Eurasia and the Han in eastern Eurasia. The Roman Empire unified Europe as never before (or since) and put an indelible cultural stamp on much of it. The empire of Han was China's formative dynasty and laid the foundation for a large and powerful state. The Han capital, then called Ch'angan (now Xian), was known as the Rome of China; Rome was the Ch'angan of the Mediterranean. The Silk Route not only transferred goods between East and West, but also tales of splendor and power.

And it was warm, in western Europe as well as in eastern Asia. The Romans planted grapevines in Britain and left behind a thriving wine industry. The agricultural frontier moved steadily northward in Scandinavia, and treelines and pastures moved upward on Alpine slopes. Meanwhile Europe's medieval cities mushroomed as architects endowed them with some of the civilization's major works: the Notre Dame Cathedral on an island in the Seine River in the heart of Paris; the cathedral at Chartres, the cathedral at Canterbury, and numerous other triumphs of Gothic styling and engineering. In China, the Tang Dynasty (618–907) brought a golden age of expansion and consolidation, architecture and art. Superbly designed pagodas heralded the diffusion of Buddhism; Xian was the cultural capital and the largest city in the world. The ensuing Song Dynasty benefited from the unprecedented agricultural productivity of the North China Plain and the rice fields to the south: by the end of the dynasty, 1279, China had an estimated 100 million inhabitants. And when the Mongols invaded China to establish their Yuan Dynasty they, too, benefited for some time from the beneficent conditions of the Medieval Optimum.

But not for long. Had the Chinese and the Europeans been exchanging weather information the way we do today, the Chinese would have been alarmed at what they heard from Europe. Because in the west, the mild and pleasant conditions prevailing for so long that they had become the accepted norm showed signs of ending. Winters got colder, May frosts, hardly known for centuries, became common. Early fall frosts led to local famines. Persistent droughts hit some parts of Europe; destructive floods struck elsewhere. Britain's wine industry was erased by cold in a matter of decades. By the turn of the fourteenth century, Alpine glaciers began to advance. Greenland's small settlement had long since disappeared; Iceland was abandoned as well. Weather extremes abounded, not only in the form of record cold snaps but also as searing summer heat and raging storms. Nature seemed to be preparing for one of those climatic reversals so common during the Pleistocene. Was the Holocene itself coming to an end?

THE LITTLE ICE AGE

To the farmers, winegrowers, and seafarers of the fourteenth century it must have seemed so. Increasing cold, decreasing rainfall, frigid winds, and shortened growing seasons made for dwindling harvests, failing farms, and seas too stormy for fishing. Famines struck all over Europe, just at a time when more people were clustered in towns than ever before. And the climatic record, pieced together from farmers' diaries (those of winegrowers are especially useful), tree ring research (dendochronology), ice cores, contemporary writings, illustrative paintings, and surviving sketches and drawings, justify the designation of the post-1300 period as a reglaciation. We now know that this return to colder times, marked by advancing mountain glaciers and thickening Subarctic ice, would end in the mid-nineteenth century, and that even the worst of it, starting in the late 1600s, did not lead to full-scale Pleistocene glaciation. Whatever was happening caused havoc in Europe, and in other parts of the world as well, but of course those who experienced it were unaware of the long-term implications. Only when new methods of analysis became available did scientists realize what had happened—and then they gave the episode an inappropriate name. This temporary cooling was no ice age: it was a minor glaciation, and not the first over the past 7,000 years. But the name "Little Ice Age" certainly was more dramatic than "Minor Glaciation," and it stuck.

To those affected, it was anything but little or minor. Europe's climate fluctuated wildly, often suddenly, so that recovery would be followed by renewed famine, as populations mushroomed and then collapsed again. In the fourteenth century, the geography of eastern and western Eurasia became fatally interlocked. The salubrious conditions that enabled Mongol peoples to thrive and expand, leading to a Mongol dynasty in China, also energized their penetration westward. In the process, Mongol migrants and their horse caravans picked up the strain of bacteria that brings on the bubonic plague, and its vector, the flea, rode into Europe on rats and people ahead of their relentless advance. The Black Death swept over an already weakened Europe in waves that often killed half the population or more, and recovery, medical as well as environmental, did not start until the last quarter of the fifteenth century.

In China, meanwhile, the full impact of the Little Ice Age occurred after the end of the Mongol (Yuan) Dynasty (1368). The early Ming rulers inherited a populous state sustained by wheat in the north and rice in the center, linked by the Grand Canal and other busy waterways. Late in the fourteenth century the Ming rulers, exhorted by the legendary admiral Chung Ho, authorized the construction of an ocean-going fleet that would stake China's claim, and enhance its reputation, in the Indian Ocean and beyond. The fleet eventually numbered more than 6,000 ships, the largest carrying as many as 500 men; these were 400 feet long, had four decks and nine masts, and carried sufficient fresh water and supplies to sail for 30 days. Nothing built in Europe even began to approach these vessels in terms of technology or capacity, and the first expedition, in 1405, involved 315 ships and 27,000 men. Later voyages reached the Persian Gulf and the Red Sea as well as East Africa, possibly as far south as Sofala. The Chinese seemed poised to round the Cape of Good Hope and enter the Atlantic.

But then disaster struck at home. The first onslaught of the Little Ice Age came later than it did in Europe, but it was no less severe. Interior rains failed, rivers dried up, the wheat crop shrank, famines broke out, and social disorder and epidemics raged. The Ming rulers ordered an end to the maritime expeditions, dictated the burning of all ocean-going vessels, and instructed the Nanjing shipyard, then by far the largest in the world, to build only barges that could navigate the Grand Canal with cargoes of rice, thus to alleviate the plight of the colder, drier north. Environment may not determine the capacities of humans, but environmental events can decisively influence the course of history.

CRISIS IN EUROPE

It is no exaggeration to say that the vicissitudes of Western Europe's Little Ice Age climate brought about what cultural geographers refer to as the second Agricultural Revolution. Farm implements were improved; field methods (planting, sowing, watering, weeding, harvesting) got better, transportation and storage of produce involved less waste and loss. New crops were tried (not always with good results); marketing in the growing urban areas became more efficient. All this was, literally, a matter of survival, because toward the end of the sixteenth century there were signs that the Little Ice Age had even worse in store. The century closed with one of the most extreme decades in Europe's known environmental history, and during the seventeenth, conditions were worsened by a series of volcanic eruptions in Southeast Asia, precipitating colder spells in an already frigid region (Grove, 1988).

Some climatologists call the period from about 1650 to 1850 the "real"

Little Ice Age, citing the environmental crisis that gripped Europe during those 200 years as much more severe than anything that had gone before. And indeed, between 1675 and 1735 the planet appears to have experienced the coldest cycle of the millennium. Growing seasons in parts of Europe were shortened by as much as six weeks; ports were blocked by ice; the Denmark Strait between Iceland and Greenland remained ice choked and impassable even during the summers. Sea ice formed and remained in place over the North Sea as far as 35 miles (55 km) from shore. Priests and their parishioners prayed at the faces of fast-advancing Alpine glaciers threatening villages and farms.

Was this indeed a global phenomenon? In his book *The Little Ice Age* archaeologist Brian Fagan describes how the Franz Josef Glacier on New Zealand's South Island "thrust downslope into the valley below, smashing into the great rainforests . . . felling giant trees like matchsticks" (Fagan, 2000). In North America, our growing understanding of the Little Ice Age helps explain why the Jamestown colony collapsed so fast, a failure attributed by historians to ineptitude, lack of preparation, and racist attitudes toward the Native Americans in the area. The chief cause probably was environmental. Geographer Don Stahle of the University of Arkansas and his team, studying tree ring records that go back eight centuries, found that the Jamestown area experienced a seven-year drought between 1606 (the year before the colony's founding) through 1612, the worst in nearly eight centuries. European colonists and Native Americans were in the same situation, and their relations worsened as they were forced to compete for dwindling food and falling water tables. The high rate of starvation was not unique to the colonists. They, and their Native American neighbors, faced the rigors of the Little Ice Age together.

In Europe, there was little respite. The decade of the 1780s brought one crisis after another. A gigantic volcanic eruption on Iceland in 1783 lasted for eight months and ejected an estimated 100 million tons of ash, sulfur dioxide, and other pollutants into the atmosphere. The Laki eruption lowered temperatures in North America by 7 degrees Fahrenheit, and brought on a series of frigid winters in Europe, Russia, and even North Africa. In February 1784, ice blocked the entire lower Rhine River, producing the worst floods in recorded history and causing food shortages and general economic distress. Violent weather in Western Europe in 1788 included hailstorms that felled forests (one report refers to hailstones 15 inches in diameter) and storms that flattened crops. The French Revolution had other causes, of course, but its timing undoubtedly was related to the food shortages recurring during that dreadful decade. Nor was Napoleon Bonaparte fortunate in his timing when he marched against Russia in 1812. The period from 1805 to 1820 was one of the coldest in the "real" Little Ice Age, and when Napoleon's armies invaded Russia their biggest adversary was the bitter winter, with which Russian forces were more familiar.

DISTANT THREAT

Just when it must have seemed that conditions could not get any tougher, they did—not because of an atmospheric event but as a result of a volcanic eruption on the other side of the planet.

On April 5, 1815, the Tambora Volcano on the island of Sumbawa in what was then the Dutch East Indies, located not far east of Bali, rumbled to life. Less than a week later it was pulverized in a series of explosions that could be heard a thousand miles away, killing all but 26 of the island's population of 12,000. When it was over, the top 4,000 feet of the volcano were gone, and most of what is now Indonesia was covered by debris. Darkness enveloped much of the colony for weeks, and tens of thousands died of famine in the months that followed. Colonial reports describe fields covered by poisonous ash and powder, waters clogged by trees and cinders, air rendered unbreathable by a fog of acid chemicals.

Tambora's explosions rocketed tens of millions of tons of ash into orbit, darkening skies around the world. What began as a narrow equatorial band of ash and dust gradually widened into a globe-girdling membrane that blocked part of the Sun's radiation. By the middle 1816, it was clear to farmers everywhere that this would be a year without summer, a growing season without growth. In Europe, food shortages were acute and grain prices rose rapidly, forcing governments to close their borders to prevent speculation. Food riots nevertheless broke out in the towns, and in the countryside armed gangs raided farms and stores. In the United States, the "year without summer" was especially difficult on the farms in New England, where corn would not ripen, grain prices escalated, and the livestock market collapsed. We can only guess at the impact of Tambora's eruption in other parts of the world, but there can be no doubt that 1816 was a desperate year in the Little Ice Age—a crisis that reminds us of the risks under which all of humanity lives.

THE HUMAN FACTOR

To put these chronological events in spatial perspective, we should remind ourselves of what was happening to the human world as the Little Ice Age came to its mid-nineteenth-century end. The Industrial Revolution was gathering steam; the colonial era was transforming societies and economies from Central America to Southeast Asia. Europeans were populating and dominating distant lands, fighting among themselves even as they exploited their imperial domains. And population growth was accelerating. Around the time of the Tambora eruption, the Earth's population was about 1 billion, perhaps twice what it was at the beginning of the Little Ice Age. Since then, a human population explosion has coincided with the post–Little Ice Age warming that has been in progress, with one major interruption from approximately 1940 to 1970, since about 1850. Consider this: by 2015, two centuries after Tambora, the Earth will carry seven times as many people as it did when that volcano exploded. How would the world cope today with a "year without summer," not to mention a Toba catastrophe—or a sudden return to Pleistocene glaciation?

GLOBAL WARMING?

Since the 1850s, when the Little Ice Age waned and a slow but more or less persistent warming phase began, geographers and other scientists have learned much about the workings of our planet, the interlocked systems and cycles that drive environmental change. Reliable predictions, however, cannot yet be made. We appear to be experiencing a warming phase similar to the Medieval Optimum, the rate of temperature increase enhanced, this time, by human activity. Some scientists argue that the current warming is not significantly greater than occurred during the Medieval Optimum, but in the absence of accurate measurements a millennium ago that argument is moot. In any case, the vast volume of gases now emanating from factories and vehicles is substantially augmenting the "greenhouse" effect of the atmosphere, and until the natural cycle reverses and Pleistocene cooling resumes, we are headed for an uncertain environmental future.

This uncertainty is rendered even more problematic when we compare what has happened so far during the Holocene and what occurred during the previous Pleistocene interglacial, the Eemian. James Hansen has pointed out that, during the Eemian, temperatures rose even more than they have during our Holocene (and human action was not a major or even minor part of the problem then). His calculations indicate that sealevel, during the Eemian, rose as much as 10 to 16 feet (3 to 5m) above the current level (Hansen, 2004). It is true that the peak temperatures during interglacials are not invariably the same: the interglacial about 320,000 years ago peaked at higher temperatures even than the Eemian, but the one 240,000 years ago never got

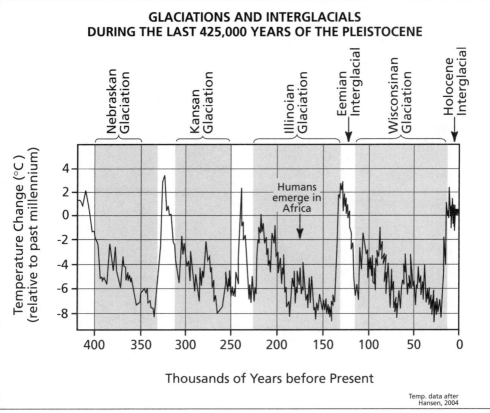

**GLACIATIONS AND INTERGLACIALS
DURING THE LAST 425,000 YEARS OF THE PLEISTOCENE**

Fig. 4-2

as warm as either the Eemian or the Holocene (Fig. 4-2). So we may be about to experience the highest temperatures of the entire Holocene—or we may be on the verge of the same precipitous cooling you see ending earlier inter-glacials in the diagram. Obviously the wild card is human action: there is no doubt that it enhances global warming, but it is far from certain to what degree; nor do we know whether human-caused enhancement of the natural atmospheric greenhouse effect might be enough to stave off a return to full-scale glaciation altogether. Somehow it seems arrogant to assume that we can overpower the grand design of nature (given what nature has overcome after Chicxulub and Toba), but some scientists actually argue that we may be heating the planet forever.

We should be aware of another caveat when we refer to "global warming." While this may be appropriate with reference to a planetary average, in other ways it is as much a misnomer as "Little Ice Age." For some regions, there will be nothing global about it.

The effects of planetary temperature increases will be regional, not

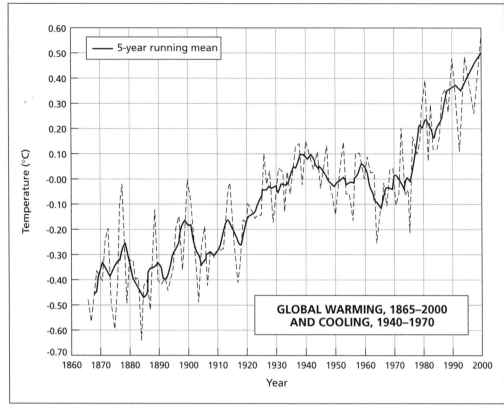

Fig. 4-3

worldwide—as the severe 2003 heat wave in Europe, which coincided with unusual cool spells in parts of North America, underscored. But in general terms, glaciers and icecaps are in retreat, agricultural frontiers are expand-ing (the British wine industry has made a comeback after nearly a thousand years), high-latitude seas are calmer and storms are fewer. At the same time, there is a troubling, rising incidence of extremes, of torrential rains in desert areas, parching droughts in humid zones, early autumn frosts and searing summer heat, intense tropical storms and early as well as late blizzards. If we could interview a climatologist from the 1300s, we might well be told that it all seems rather familiar. Climatic reversals seem to be presaged by such exceptional weather events.

It is tempting to call the present, post-1850 warming phase the Indus-trial Optimum given the role of anthropogenic (human-source) greenhouse gases. The warming trend was interrupted between 1940 and 1970 just when the Industrial Revolution was in full gear, and the scientific as well as the popular literature of the 1950s and 1960s was full of forecasts of impending glaciation (Fig. 4-3). I was a graduate student at Northwestern University in the late 1950s, and "global cooling" was the mantra in those days just as "global warming" is today. Scientists have since proposed theories explaining why three decades of global cooling could have occurred just when the industrializing Soviet Union's prodigious output of greenhouse gases was being added to the global volume, but whatever the reason, the event should give us pause. In the United States, many people responded by moving southward (the "Sunbelt" notion was thus born). Here was a hint of nature's power to override human actions.

Will we, in our enormous numbers, ever be able to adapt to rapid environmental change and save ourselves from the chaos that attended the onset of the Little Ice Age? As Brian Fagan states, it may be an illusion to suggest that humanity, through its technological prowess, will ever be able to adjust to the kinds of changes nature has on the record. "Climate change," he writes, "is almost always abrupt, shifting rapidly within decades, even years . . . it is unpredictable, and sometimes vicious. The future promises violent change on a local and global scale . . . such cycles are frightening to contemplate in an overpopulated and heavily industrialized world" (Fagan, 2000).

Scientists are zeroing in on the question of abrupt climate change, and they are finding that the current global warming is likely to trigger rapid environmental shifts that could cause chaos on the planet. Ice cores from Greenland, nearly two miles long and revealing a climatic record spanning the past 110,000 years, record wild fluctuations including slow warming and sudden cooling, the latter (before the Little Ice Age) sending icebergs as far south as the coast of Portugal (Alley, 2002). High-latitude cold had regional effects all over the world. For example, about 5,000 years ago a northern cold spell coincided with the drying of the Sahara astride the tropic, converting it from a verdant landscape with rivers and lakes to the parched, rocky, and sandy wasteland it is today. That desiccation did not happen over millennia—it occurred over just a few decades, with far-reaching impact on the entire continent's human geography. The Greenland ice cores show that nature's variability is endless and restless. Now, scientists theorize, the additional impact of human activity on the global atmosphere may trigger even more sudden climate change than prevailed long ago. Sooner or later we will face extremes that come upon us quickly and will give us little time to find ways to cope with the consequences. For all our technological prowess, we still depend on nature to sustain us.

We are living, in the words of the geneticist Stephen Oppenheimer, in the autumn of the interglacial Holocene, the brief epoch that has witnessed the transformation of our human world from small villages to megacities and from simple to complex cultures, from isolated communities to intercon-

nected empires and from stone tools to spacecraft. On the scale of our life-
times, it has all happened in the past few seconds, and our modern experience
with the kinds of challenges nature can pose—from meteor impacts to glaciations—is virtually nil. We will never be able to control climate change, but we
may be able to mitigate it somewhat by limiting our greenhouse-gas emanations. And we should begin planning for worldwide coordination in the event
of global natural emergencies caused by nature, of which we already have been
amply warned. The impact of an episode of rapid climate change poses as
great a potential challenge to this nation as any it will face in the years ahead.

CLIMATE AND WEATHER ON THE MAP

In this chapter we have discussed climate change, but most of us are concerned on a daily basis with the weather. The difference is obvious: the term
"climate" represents an overall picture of the year-round conditions prevailing at a given place. It is an average of all available data, especially temperature and precipitation and their seasonal ups and downs. That's why we
speak of a tropical climate or an Arctic climate. The *weather*, on the other
hand, refers to the conditions at a given moment. If it is 90 degrees Fahrenheit and raining somewhere with high humidity and little wind, that is the
weather—probably somewhere in a tropical climate.

All the weather-making factors combined—the sweep of the Sun's energy,
the rotation of the Earth, the circulation systems of the oceans, the movement of pressure systems, the rush of air in jet streams—produce a global
pattern of climates that may look complicated at first but is actually remarkably simple. This is one of those maps that is worth a million words, because
it allows us, at a glance, to determine what the prevailing climate is anywhere on Earth, and what we may expect in the way of weather under those
climatic conditions (Fig. 4-4).

We owe this remarkable map to the work of Vladimir Köppen (1846–
1940), who devised a scheme to classify the world's climates based on indices
of temperature and precipitation. Since his time, there have been many efforts to improve on Köppen's classification and regionalization of climates,
but nearly a century after its creation, and despite the greater accuracy of
climatic data today, it has stood the test of time.

Using not only climatic but also biotic information, Köppen established
six major types of climate. He used letters to identify these on his map:

A: Equatorial, Tropical, Moist
B: Desert, Dry

C: Midlatitude, Mild
D: Continental, Harsh
E: Polar, Frigid
H: Highland

Köppen added other letters to provide more detail within each of these climatic regions, but even without these, his map is very useful. Take, for example, the climate prevailing over most of the southeastern United States, Cf on the map. If you're familiar with that climate and its local weather (for example, in Atlanta, or Nashville, or Charlotte), you'll have a pretty good idea of what it's like in much of eastern China, in southeastern Australia, and in a large part of southeastern South America. Chicago's climate is a Df climate, with colder winters and hot summers; you'd experience something close to it in Moscow and Sapporo (northern Japan). The map shows you instantly where the "real" rainforest climate prevails, where the world's deserts are, and where the best grape-growing areas lie. Note that the Cs climate, a mild climate with a dry summer (which is denoted by the small letter s), occurs not only around the Mediterranean Sea (and thus in the famed wine countries of France, Italy, and Spain), but also in California, Chile, South Africa, and Australia. So you know what to expect, weatherwise, in Rome, San Francisco, Santiago, Cape Town, and Adelaide.

Of course it's possible to take part of this world map and draw it at a much larger scale. That takes away our ability to make worldwide comparisons, but it increases the amount of information that can be provided for a smaller region. Travel kits almost never include such a detailed climate map. They should.

Ever since Köppen published his map of world climates, geographers and others have been intrigued by the apparent spatial relationship between certain climates and certain successful, powerful societies. Equatorial and tropical climates, apparently, do not favor the countries over which they prevail; none of the world's major powers, present or recent, lie here. Desert climates also do not appear to be conducive to big-time status. The same seems true of high-latitude, polar climates.

Are the mild midlatitude and the harsh continental climes the best our planet has to offer? Ellsworth Huntington had no doubt about it. "The people of the cyclonic regions," he wrote in 1942, meaning the midlatitude cyclonic zones, "rank so far above those of other parts of the world that they are the natural leaders . . . [they lead in terms of productivity, but] their greatest products are ideas and the institutions to which these give rise. The fundamental gift of the cyclonic regions is mental activity" (Huntington, 1942).

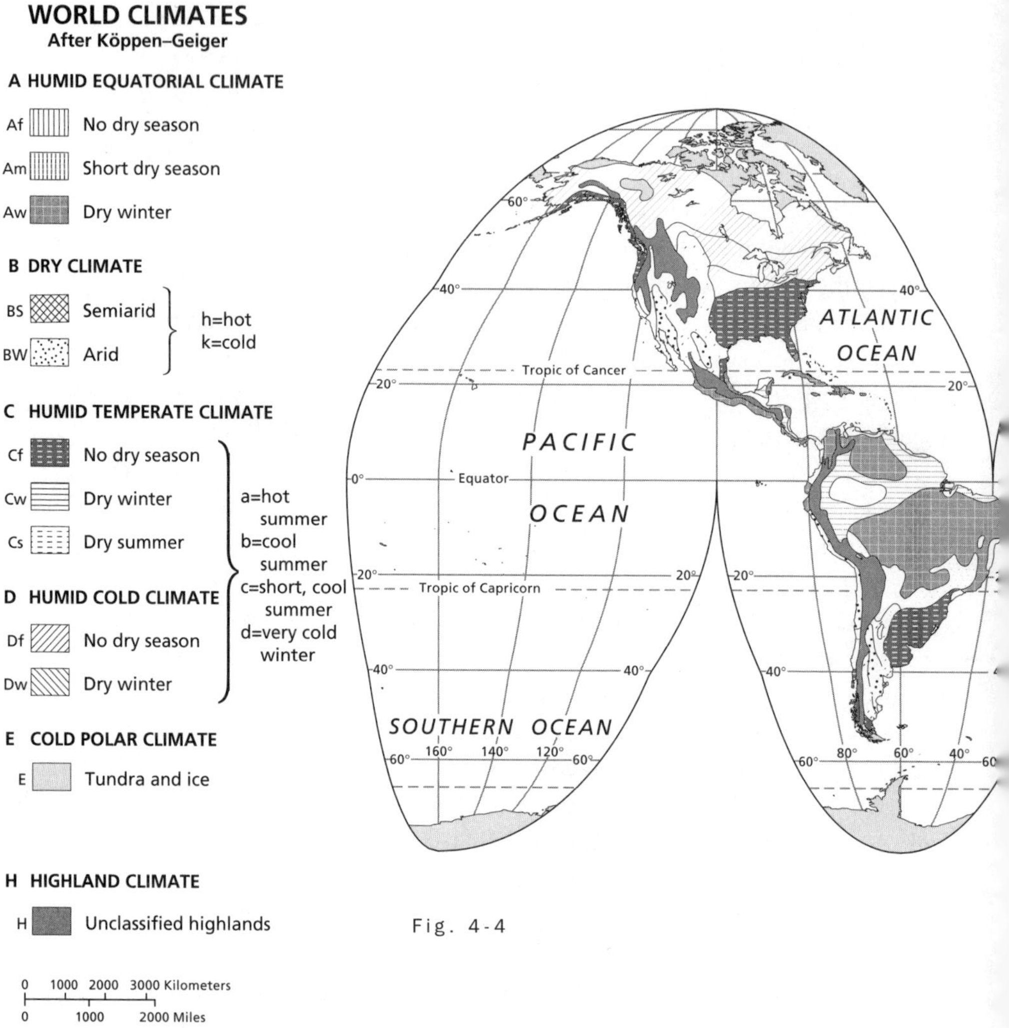
WORLD CLIMATES
After Köppen–Geiger
A HUMID EQUATORIAL CLIMATE
Af No dry season
Am Short dry season
Aw Dry winter
B DRY CLIMATE
BS Semiarid
BW Arid
h=hot
k=cold
C HUMID TEMPERATE CLIMATE
Cf No dry season
Cw Dry winter
Cs Dry summer
a=hot summer
b=cool summer
c=short, cool summer
d=very cold winter
D HUMID COLD CLIMATE
Df No dry season
Dw Dry winter
E COLD POLAR CLIMATE
E Tundra and ice
H HIGHLAND CLIMATE
H Unclassified highlands
0 1000 2000 3000 Kilometers
0 1000 2000 Miles
ATLANTIC OCEAN
PACIFIC OCEAN
SOUTHERN OCEAN
Tropic of Cancer
Equator
Tropic of Capricorn

Fig. 4-4

Such observations made Huntington famous; they also got him into a lot of trouble. By extension, it's possible to conclude that peoples living long term under cyclonic conditions would be intellectually superior to others, and from this, it was but a short step to Nazi master-race philosophies. Environmental determinism, the idea that environment (chiefly climate) controls the capacities and destinies of societies, quickly got a bad name.

That was unfortunate, because Huntington had raised one of geography's central questions, and in the context of his time, he made enormous contributions to our understanding of climatic change and its impact on human societies. These days it is fashionable to join the chorus of disparagement,

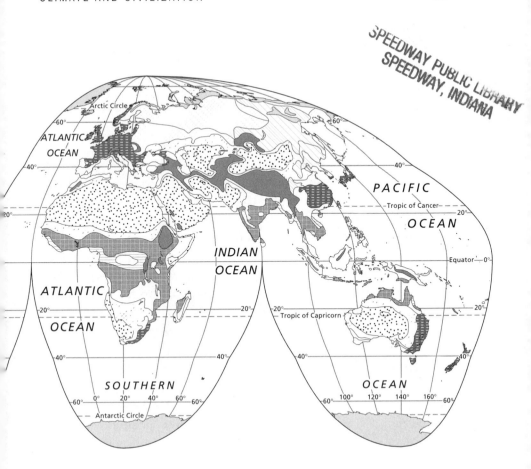

but many who do so have never read Huntington's last major (and monumental) work, *Mainsprings of Civilization*, which appeared in 1945, two years before his death. *Time* magazine, in its review of this book, proclaimed it comparable to one of the greatest of historical works, Arnold Toynbee's 12-volume *Study of History*.

Huntington argued that the seasonality of midlatitude climates had favored peoples who remained for many generations under these demanding, yet stimulating regimes. He explained the rise and fall of societies in terms of the "sweep" of climate change. Köppen's map, he frequently said in his lectures, should be seen as a still photo of a changing Earth—it represents the way things are today, but not as they were yesterday, nor as they will be tomorrow. When the Maya civilization rose to greatness, when West African states arose and prospered, when Muslim civilization flourished and outshone its European contemporaries, it was because climate induced it. "The well-known contrast between the energetic people of the most progressive parts of the temperate world and the inert inhabitants of the tropics," he wrote, "is largely due to climate."

Small wonder that environmental determinism suffered the fate it did. Still, in recent years, the notion has been rediscovered and subjected to more rigorous analysis. Jared Diamond, who is on the faculty of the Department of Geography at UCLA, in his remarkable work *Guns, Germs, and Steel*, argues that it is not climate alone, but the environmental opportunities offered by a combination of natural conditions ranging from wildlife to plants and from water supply to relief that put certain peoples at an advantage over others (Diamond, 1997). Lose those opportunities, and progress is halted. Benefit from them for a long time, and advantage endures.

On one point all can agree, and it was one Huntington made time and again: climate is cyclic, and the Köppen map represents the present, not the past nor the future. In this time of global warming, A climates are gaining on C climates, C climates are shifting poleward into D climates, and in the Arctic and Antarctic, melting ice and warming water are altering natural habitats (there is even talk of the opening of a maritime northern passage from the North Atlantic to the Pacific). It has all happened before—during the previous interglacial, the Eemian, it got so warm that sealevel reached about 15 feet (4 meters) higher than it stands today. But the Eemian ended with a glacial bang, a return to glacial times so sudden that our ancestors emigrating from Africa froze in the Levant. What will the next millennium bring?

5

A FUTURE GEOGRAPHY
OF HUMAN POPULATION

When I lived and worked in Washington, D.C., my weekly schedule often involved trips to nearby towns like Alexandria, Silver Spring, Reston, and Vienna. It was almost never easy. Year by year, the roads seemed to become more crowded. I learned to leave extra early for appointments at the University of Maryland or George Mason University because after the slow ride there would be a long search for a parking place. I remember one time, finding myself trapped in one of Northern Virginia's traffic jams with cars in front, behind, and on both sides of mine, when the traffic stopped dead. Lines in all lanes stretched as far as the eye could see, in both directions. It was a cool day, and most drivers had their windows down. A fellow in a Mercedes next to me leaned over.

"Must be the population explosion!" he shouted. "It's hit us, and we'll all expire!" He must have got a noseful of the exhaust fumes now swirling around, because he raised his windows and turned on his air conditioning. So did almost everyone else: the lineup of idling engines created a soupy atmosphere on the nearly windless afternoon.

My neighbor was joking, but he had a point. If there is anyplace in America where you can experience for yourself the impact of rapid population growth, it's in the metropolitan areas, especially in the suburbs. Where I sat waiting for the traffic to move, forests of newly built townhouses stretched in all directions, having gobbled up farmlands and meadows. Ugly pylons and wires carried electricity to the new developments. Construction crews were at work on roads, sidewalks, drains. Trucks lined up to join the highway traffic, which moved at a snail's pace if at all.

"Is it like this every day?" I asked a gas station attendant up the road.

"It's usually worse," he said. "It's pretty quiet today, people aren't even honking their horns. I'm from the Los Angeles area, and to me, this is nothing."

Being caught in slow-moving traffic is as good a time as any to reflect on a concern that is a topic of conversation wherever you go in the world: population growth and its attendant problems. People are apt to attribute most of the world's ills, from persistent malnutrition in parts of Africa to war in the Middle East, and from unwanted immigrants to crowded highways, to over-population. You hear this refrain even in countries where the population is stagnant or actually declining, such as Japan, Russia, and Italy. But those who live in Western or Westernized societies and complain about problems arising from population growth simply haven't experienced the real conse-quences in such places as India or Nigeria. Spend some time in Kolkata or Lagos and you will soon forget the discomforts of Northern Virginia or sub-urban Los Angeles.

THE GLOBAL SPIRAL

The explosion of global population during the past two centuries, and espe-cially during the twentieth century, is common knowledge now. About 200 years ago, shortly after the onset of the Industrial Revolution, the Earth's total population was only about 900 million, which is less than the present-day population of India alone. In 1820 the milestone of 1 billion was reached, and this number took more than a century, 110 years to be exact, to double. Then the upward spiral really took off. Only 45 years later (1975), 2 billion doubled to 4 billion, and at present growth rates, this number will again double, to 8 billion, around the year 2035. If there is anything to be opti-mistic about in these data, it is that the time it takes for the world's human population to double has lengthened from 45 to 60 years.

There are times when scary statistics are suspect, but not in this case. For thousands of years of human evolution and civilization's development, our numbers remained small, to be measured in tens, then in hundreds of mil-lions. Even as recently as 1650, the world's human population was about 500 million—as many as are being added now every seven years or so.

Why is this happening? A combination of factors caused what is, in truth, explosive population growth. A population's natural increase over a given period is measured, simply, by subtracting deaths from births. When the number of deaths per 1,000 people in the population is close to the num-ber of births, that population will grow very slowly. But when the number of deaths declines, the gap widens and population growth accelerates. Tak-ing the world as a whole, death rates began to decline during the eighteenth century, while birth rates remained high. That created a widening gap, and while birth rates also started to decline in some parts of the world, this de-cline lagged far behind the death rate. The lag is still in effect.

Death rates declined because of major progress in hygiene and medicine associated with the Industrial Revolution and because of their export from the source (mainly Europe) to the rest of the world. Just two inventions, effective soap and the toilet, contributed enormously to this; ever-better medicines had great impact as well. Refrigeration, water purification, and other advances also helped lower the death rate. Such progress ensured the survival of countless babies who would otherwise have died at or soon after birth. The death rate of a population reflects not only those who die after a substantial lifetime, but also those who die at birth or in infancy. Today, one of the most telling statistics reflecting a country's overall condition is its infant mortality rate, the number of babies that die within the first year of life, usually reported per 1,000. In 2004, some countries reported all-time lows of 3 (Japan, Sweden) but 18 African and several Asian countries' infant mortality rates still exceeded 100.

NATURAL INCREASE IN REGIONAL PERSPECTIVE

Enormous differences, therefore, still exist in the well-being of this planet's regional populations, as reflected not only by infant mortality statistics but by other data as well. In the mid-twentieth century, when Paul Ehrlich and others raised the "population explosion" alarm and forecast vast famines afflicting billions across the planet, the global nature of the coming crisis tended to obscure regional contrasts, although these were identifiable even then. Today, the regional geography of population change is a key concern.

At issue is the geographically variable rate of natural increase (or decrease). Populations also grow and decline through net immigration or emigration, but natural increase is what matters here. A map of the world showing the countries where population is either stagnant (0.0 percent growth rate) or declining shows a large swath of states from Japan to Sweden in this condition (Fig. 5-1). A half century ago a movement called ZPG (Zero Population Growth) called for this then-unimaginable goal as a global objective. It is now here, in regional dimensions, and it is fraught with its own problems. Countries with stagnant or declining populations face economic problems the ZPG crowd didn't think of.

Regional rates of population change vary from Russia, decreasing at 0.7 percent annually and Europe (-0.1 percent) to Subsaharan Africa, growing at 2.2 percent despite its terrible AIDS affliction. Middle America, led by Mexico at 2.1 percent, is the next-fastest-growing realm, followed closely by the Islamic world (1.9 percent). Even South and Southeast Asia grow more slowly than this. In North America, the United States has been growing at 0.6 percent and Canada at 0.4 percent.

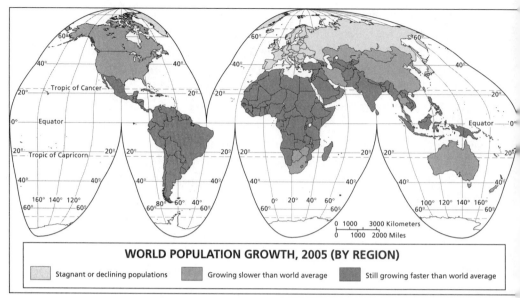

WORLD POPULATION GROWTH, 2005 (BY REGION)

Stagnant or declining populations   Growing slower than world average   Still growing faster than world average

Fig. 5-1

It is important to look at the data for individual countries, however, because even within the world's dozen or so geographic realms there are regions and countries that defy the overall trend. In the Islamic world, for example, Saudi Arabia and Yemen grow much faster than the regional average, but Iran and Tunisia grow much slower. Still, all but one country in the Islamic realm grow faster than 1.0 percent; all but one country in Europe grow slower than 1.0 percent—and the exception is Muslim Albania.

It would seem logical to conclude that religion has much to do with these figures. The Roman Catholic Church is strong in Middle America, and the Vatican has sought common cause with Muslim clerics on population issues at international meetings. But it is telling that population decline is most evident in the very heartland of Catholicism, even in Italy itself. Low rates of population increase have more to do with levels of development and with the status of women in society than with the strictures of various faiths. Another factor is democracy. Efforts to slow population growth by coercion in India led to massive resistance and voter retaliation. In communist China, an authoritarian government could decree a "One-Child Only" policy and generally make it stick. Today, India is still growing at 1.7 percent; China's rate is down to 0.7, almost on a par with the United States and Australia. But even at a low 0.7 percent, China is adding between 9 and 10 million people to its realm's population—equivalent to a Sweden—every year.

THE CONSEQUENCES FOR EUROPE

The United States and Canada are in the fortuitous position to have a still-healthy rate of natural increase that, coupled with substantial legal immigration, is sustaining economic growth and optimistic demographic projections. But Europe faces trouble. According to a recent UN study, the Earth's total population by 2050 will approach 9 billion. The (anticipated) 27 countries of the European Union, however, will see their populations decline from 482 million to 454 million in 2050. Italy may drop from nearly 58 million to 45; Germany from 82 to 69 and, if things do not change, to a mere 25 million by 2100. As Charlemagne writes in the July 19, 2003, issue of the *Economist,* "Combine a shrinking population with rising life expectancy, and the economic and political consequences are alarming. In Europe there are currently 35 people of pensionable age per 100 people of working age. By 2050, on present demographic trends, there will be 75 pensioners for every 100 workers; in Spain and Italy the ratio of pensioners to workers is projected to be 1:1." Because pensions are paid out of tax revenues, taxes will have to rise sharply to fund the generous pensions Europeans are accustomed to. Workers will demand that taxes be kept down, and labor unrest will become even more endemic in Europe than it already is. Further, countries with fairly stable populations, such as the Netherlands and the United Kingdom, will resent being enmeshed in the financial problems of other EU countries, creating potential schisms in the European Union.

One answer, it would seem, is more immigration and naturalization, but this creates its own social, political, and even economic problems. Estimates suggest that immigration (which is coming mainly from Muslim North African countries and Turkey) would have to increase five- to ten-fold to compensate for the rate of natural decrease prevailing in Europe today. The associated social strains would be unmanageable. As Europe's population shrinks and its proportion of the world population declines, dreams of an economic superpower fade. The implications of Europe's demographic data are far reaching indeed.

PATTERNS OF THE FUTURE

If we divide the world in a very general way into rich and poor, the populations of the rich countries today are growing at 0.25 percent annually and the poor countries at 1.46 percent annually (Cohen, 2003). The populations of the least-developed areas of the world, the 50 countries with the nearly 700 million poorest inhabitants, are growing by more than 2.4 percent annually. By the middle of the century, various agencies estimate, the world's overall population growth will be down to an encouraging 0.33 percent, but by then the poorest countries will still be growing at 0.4 percent while the rich countries will be shrinking at a still-accelerating 3.14 percent, further widening the quantitative gap between the well-off and the rest.

Where will the bulk of this poorer majority reside? In the burgeoning cities that already mark the "developing" regions of today. Cohen reports that, of the projected 2.2 billion increase in population between 2000 and 2030, all but 100 million will crowd into cities, so that urban areas will grow at about twice the rate of global population increase. In his just-cited article, Cohen reminds us that the rural population of the rich countries peaked around 1950, after which it declined everywhere, making "rich" countries synonymous with "urbanized" countries. The rural population of the presently poor countries is expected to peak around 2025 and will then begin a similar decline. Projections suggest that urbanization in the poor parts of the world will rise from 40 percent in 2000 to 56 percent in 2030, which is about where the rich countries were in 1950. But the numbers of residents marking the burgeoning cities of the poorer world will dwarf anything seen in the rich world during its heyday. Conurbations of 50 million or more will anchor regions of India (which is likely to become the world's most populous country) and China. As much as 80 percent of the populations of some now-poor countries will crowd into their largest cities. Already, the world's population is more than 50 percent urbanized today; urban life will be the norm for the great majority in the future.

In the process, many things will change worldwide: family relationships (proportionately more children born out of wedlock), the proportion of youngsters in the population (shrinking and destined to be exceeded by oldsters), life expectancies (longer), schooling and literacy rates (higher), and secular lifestyles (more common). The unanswerable question is how the countries of the richer world will relate to the overwhelming numbers and growing economic and military power of the currently poor.

WILL THE WORLD'S POPULATION STABILIZE?

Given the projections just cited, it is not surprising that some geographers who study population questions have been tempted to extrapolate to the end of the curve, that is, to the point where global population will stop growing altogether. If the rate of global population growth is now declining past 1.4 percent, is likely to reach 0.4 percent by 2050, and continue downward, isn't it logical to conclude that the world is going the way of Europe, reaching a stagnant population and even starting a decline?

You could ask the same question about individual countries. If Japan, after "exploding" to 127 billion in 2004, is now starting a decline, what about the United States, Mexico, Brazil, or India?

It is clear from the various projections published by agencies and scholars that these remain elusive questions. A 1995 estimate suggested that China would stabilize at 1.4 billion by 2090, but China is already well past 1.3 billion today and is likely to reach that total in about eight years, not 80. Today China's ZPG moment is still foreseen as coming before the end of the century, but at 1.9 billion, not 1.4. That means that India's stabilization level will be closer to 2 billion than the 1.6 billion recently projected for 2050. As to the United States, estimates have risen from 276 million in 2035 (already exceeded) to 480 million by the end of the century (including immigration). Obviously all these projections need to be taken with more than one grain of salt, but the fact remains: barring some unanticipated upturn in fertility, the twentieth century's population explosion may be followed, if not by a twenty-first-century implosion, then by a stabilization no one in the mid-twentieth century even contemplated. What this may mean for the future of the planet is similarly beyond informed conjecture.

THE GLOBAL POPULATION MAP TODAY

When it comes to depicting the current world population on a map, it is well to remember that no single map can adequately represent the complexities involved. At the global scale, the best we can hope for is an instructive impression. Population can be mapped in terms of density, that is, the number of people per unit area (but for a page-size map, that unit would have to be pretty large), or in terms of general distribution, using the dot method. This is the method used in Figure 5-2, where one dot represents 100,000 people. The downside of a dot map, of course, is that the dots not only coalesce but overlap in certain places, for example large metropolitan areas.

Still, this map yields valuable insights. It is a powerful reminder that, on the approximately 30 percent of the planet that is land (some of it under ice), populations continue to cluster in relatively well-defined areas, reflecting the fact that of this 30 percent, two-thirds is arid, frigid, mountainous, or otherwise inhospitable to large human numbers. Of the three greatest concentrations of humanity, all of which lie on the Eurasian landmass, two—East Asia and South Asia—lie centered on ancient river basins whose fertile soils and ample drainage supported the earliest population explosions millennia ago. China and India are the modern heirs to this ancient drama, and rivers such as the Yellow (Huang) and Yangzi in China and the Ganges (Ganga) in

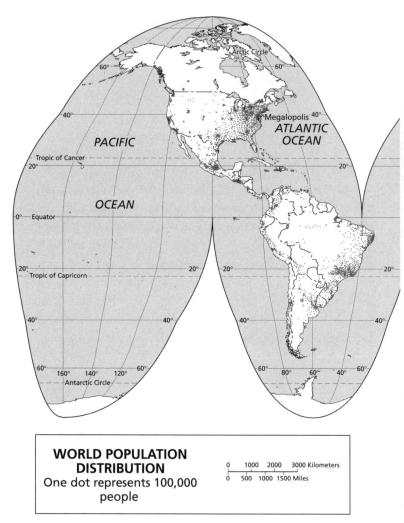

**WORLD POPULATION
DISTRIBUTION**
One dot represents 100,000
people

| 0 | 1000 | 2000 | 3000 Kilometers |
| 0 | 500 | 1000 | 1500 Miles |

Fig. 5-2

India are the arteries of human civilization, just as the Tigris-Euphrates and
the Nile were in less-crowded Southwest Asia. In these historic regions, the
majority of people still live and subsist on the land, but the inexorable migration to the cities grows and an urbanized era is in the offing. In 2004, China
reported that nearly 40 percent of its people now live in cities, and India
nearly 30 percent. But remember the sizes of these two giants' populations: it
means that more than 500 million Chinese now live in cities and towns, and
more than 300 million Indians.

The third Eurasian population cluster readily visible on the map is Europe, petering out eastward into Russia. This population, even including
some still quite rural countries, is three-quarters urbanized, and some Eu-

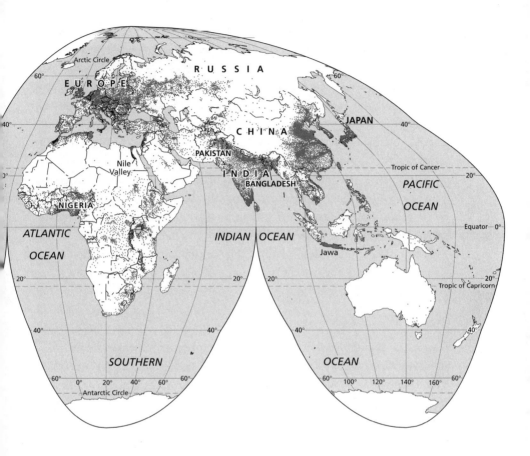

ropean countries are more than 90 percent urban. Here the transition from mostly rural to dominantly urban is nearly complete, and familiarity with urban life is a significant element in Europe's political convergence.

What is also remarkable about the distribution of population in Eurasia is the vast area of relatively sparsely populated, even open space still extant. Even China, synonymous with burgeoning population and crowded urban areas as well as rural areas, has a vast nearly empty interior. Russia's Siberia and Central Asia's steppes east of the Caspian Sea also remind us of the natural limits of settlement.

North America's eastern population cluster, centered on the coalesced cities of the east called Megalopolis by the geographer Jean Gottmann, is far smaller than its Eurasian counterparts, and South America is more sparsely peopled still. At the turn of the century Mexico City and São Paulo ranked, with Tokyo, among the world's largest conurbations, but urban agglomerations far larger than these are emerging in East and South Asia. Africa, too, is sparsely peopled for so large a landmass—there are more people in India alone than there are all of the countries of Africa—and as the map shows, the major clusters of African population lie in West Africa, centered on Nigeria, and in East Africa, focused on the Great Lakes. The Sahara is Africa's great population void, but large parts of the landmass elsewhere remain comparatively empty as well. As for the continent of Australia, the mapping method on display here truly shows how empty it remains, a function of the aridity of its "outback."

The clusters of population seen on Figure 5-2 are the repositories of the civilizations of modern humanity, the civilizations Samuel Huntington argued would find themselves in twenty-first-century conflict (Huntington, 1996). For the moment, we confine ourselves to other implications of a map that provides ammunition for those who argue that overpopulation lies at the root of most of the world's troubles.

POPULATION AND POLITICS

Obviously population growth is a key issue in environmental context. Before we assume that urbanization will improve conditions in the countryside (for example by slowing deforestation) through out-migration, we should recognize that burgeoning urban centers place enormous demands on rural areas in the form of food, water, and resources, and that the pollution emanating from huge industrializing metropolitan areas far exceeds what an equivalent number of people in the countryside would generate. People who move to the cities tend to start favoring more varied diets including more meats and poultry, resulting in deforestation in the countryside to make way for pastures.

The United Nations organizes frequent conferences on environmental issues, and I remember one in particular, the 1992 UN Conference on Environment and Development in Rio de Janeiro. When I saw the agenda, I was surprised at the lack of attention given to the population issue. When I got the roster of participants (delegations consisted of political representatives, scholars, and hangers-on) I was less surprised: few if any were geographers. The population issue was not a salient topic, but it should have been. There's little point in making commitments to protect what remains of global natural environments without taking demographics into account.

The United States delegation had an uneasy time at this conference, and the president, George H. W. Bush, came under some severe criticism for not providing sufficiently strong leadership on environmental issues. But there was plenty of censure to go around. One reason so little was said about population policies had to do with the fact that the Roman Catholic Church was still very powerful in Brazil (that power has waned in recent years as evangelical churches have made inroads). Pope John Paul II had made a visit to Colombia a few years earlier and reiterated his opposition to artificial birth control, in effect exhorting Colombians, and tens of millions of listeners throughout the realm, to have as many children as they wished. This was not a good place or time to promote practices that would have a more salutary effect on future global environments than all the touted pollution-reduction programs combined.

This does not mean that United States policy on population has been focused or steady. Successive administrations have alternately helped and hindered family-planning programs here and elsewhere.

And yet, what is happening to population growth worldwide should be of direct concern to all Americans. Underdeveloped countries coping with high growth rates see their hopes of improved living standards dashed and their dependence on foreign help raised. Overpopulation leads to economic conditions that generate desperate emigrations—and the migrants often cross American borders. Mushrooming populations stress the capacities of governments to exercise control—and American relief efforts plus, as in Somalia, policing campaigns cost lives and money. Lately, the American public has begun to worry about these issues, and the immigration question is moving higher up on the national agenda. The overwhelming approval of Proposition 187 in California during the November 1994 elections reflected a rising anti-immigrant feeling—in a nation forged of immigrants. More recently, we have seen television coverage of illegal border crossings and reports on efforts to erect physical barriers to stem the tide of illegal immigration from Mexico; CNN has almost daily news reports under the rubric "broken borders." But on the question of assisting other countries in family-planning campaigns, there is no consensus in the United States.

POPULATION AND ENVIRONMENT

Even if human population stabilizes some time in the second half of this century, and even if it then commences an overall decline, it will be too late for much of what remains today of forests and wildlife that sustain environments and link us with our past. Biologists estimate that there may be as many as 25 million types of organisms on Earth, perhaps even more; most have yet to be identified, classified, or studied. *Homo sapiens* is only one of these, and in ten millennia our species has developed a complex culture that is transmitted from generation to generation by learning and is to some degree encoded in our genes. We are not unique in possessing a culture: gorillas, chimpanzees, and dolphins have cultures, too. But ours is the only species with a vast and complex array of artifacts, technologies, laws, and belief systems.

No species, not even the powerful dinosaurs of epochs past, has ever affected earthly environments as strongly as humans do today. The dinosaurs, and many other species, were extinguished by an asteroid impact. Some biogeographers see an analogy and suggest that the next great extinction may be in the offing, caused not by asteroids but by humans, whose numbers and demands are destroying millions of species—and with them the inherited biodiversity of this planet.

This destructiveness is not just a matter of modern technology and its capacity to do unprecedented damage, whether by wartime forest defoliation, peacetime oil spills, or other means. Human destructiveness manifested itself very early, when fires were set to kill whole herds of reindeer and bison, and entire species of large mammals were hunted to extinction by surprisingly few humans. The Maori, who arrived in New Zealand not much more than 1,000 years ago, inflicted massive destruction on the native species of animals and plants in their island habitat, long before modern technology developed more efficient means of extinction. Elsewhere in the Pacific realm, Polynesians reduced the forest cover to brush and, with their penchant for wearing feather robes, had exterminated more than 80 percent of the regional bird species by the time the first Europeans arrived. The Europeans proceeded to ravage species ranging from snakes to leopards. Traditional as well as modern societies have had devastating impacts on their ecologies, and on the ecologies of areas into which they migrated.

Is wanton destruction of life a part of human nature, whatever a society's cultural roots? The question is as sensitive as questions about racism and sexism. Still, regional differences in attitude and behavior can be discerned. African traditional societies hunted for food or for ceremonial reasons, but not for entertainment or amusement. The notion of killing for fun and fashion was introduced by Europeans. Hindu society and religious culture in India are more protective of the natural world than many others. The extermination and near extermination of many species of animals in India took place during (Muslim) Moghul and European colonial times.

Malevolent destruction of the environment continues in various—indeed many—forms today, ranging from the deliberate spilling of oil and setting of oil fires by Iraqis during the 1991 conflict over Kuwait, to the mercury poisoning of Amazonian streams by Brazilian gold miners. For the first time in human history, however, the combined impacts of humanity's destructive and exploitative actions are threatening the entire Earth's biodiversity. Most of that biodiversity has been concentrated in, and protected by, the great equatorial and tropical rainforests of South America, Africa, and Southeast Asia. Now the onslaught on this last biogeographical frontier is under way, and for the future of the planet, the consequences may be catastrophic.

THE PENALTY OF POVERTY

How is it that the poorest of the world's countries tend to have the highest rates of population growth? Neither poverty nor hunger seems to slow rates of natural increase in such countries as Ethiopia, Nigeria, or Bangladesh. For a period during the third quarter of the twentieth century, Kenya had a population growth rate of 4 percent—the highest in the world as Kenyan women had seven, eight, even nine children. As a geographic realm, Subsaharan Africa grew faster than any other in the world. This reflected Africa's rural character and the strength of its cultural traditions: having many children was a matter of family security and status. Several children would die early, the father wanted sons, and a high number of births improved the odds. Infant and child mortality (the number of offspring who die before their fifth birthdays) in poor countries is horribly high, even today, but they're lower than they were during the past century. Combine the continued fertility with declining mortality, and you get skyrocketing growth rates.

And you get vulnerability to disease among the survivors. In the last quarter of the twentieth century the affliction called AIDS—Acquired Immune Deficiency Syndrome—appeared in Africa and spread worldwide. By 2004, according to UN data, more than 40 million people were infected, 33 million in Subsaharan Africa. In 2003, the latest year for which information is available, 2.6 million people died of AIDS, 2 million of them in Africa. AIDS was spreading in Russia, India, and China, but the hardest-hit realm of the world was also the poorest—Subsaharan Africa.

As the number of deaths from AIDS rises, African countries' vital statistics show the demographic impact. In hard-hit South Africa, where 20 percent of all persons aged 15 to 49 may be infected, life expectancy declined from 66 to 51 in just ten years. In neighboring Botswana it fell from 60 to 39 between 1994 and 2004, the latter a figure not seen since the Middle Ages. In Zimbabwe, population growth dropped from 3.3 percent to 0.9 percent. Some demographers are projecting population declines of 10 to 20 percent in the worst-afflicted countries. Meanwhile, AIDS is killing parents in Africa at such a rate that, by 2010, it will have an estimated 20 million orphans.

AIDS is battering Africa's already weak economies. Ordinary Africans

cannot afford the treatments available to the sick in the richer countries
of the world, and tens of billions of dollars are needed for a coordinated,
continent-wide attack on the pandemic. President Thabo Mbeki of South
Africa angered much of the world by arguing that AIDS has its origins in
poverty, not the human immunodeficiency virus (HIV), and while he was
wrong on the science he was right on the social front: poverty condemns
tens of millions of Africans to inadequate treatment or none at all. This
is yet another manifestation of aggregate misfortune that has plagued Africa—more than any other world geographic realm—for millennia.

And let us not forget that hundreds of millions of people (825 million,
by one recent UN estimate, a majority of them children) still suffer from
malnutrition or worse in these early years of the twenty-first century. They
pay the price for being in the wrong place at the wrong time, because the
truth is that all people on Earth today could be fed—not well, but adequately
for survival—if there were ways to distribute food to them and to make it
affordable for them. Tragically, that is still not happening. True, many millions who live in remote isolation and depend almost totally on what they
grow and raise are subject to the vagaries of their environment and suffer
when droughts or other natural disasters strike. But many more millions go
hungry because of the failure of governments to achieve stability and security. Decades of war in equatorial Africa have caused dislocation on a massive scale and starvation across a region where subsistence is the way of life.
To the north, the government of Sudan has used food as a weapon, denying it
to millions of refugees from a decades-long war between the Muslim regime
in Khartoum and the African minority in the far south. When a settlement
to this conflict, having taken its dreadful toll, was signed in January 2005, a
new humanitarian crisis arose in the country's west. While the UN debated
terminology (could this be defined as genocide?) thousands of villages in
Darfur Province were burned, hundreds of thousands of Africans were dislocated, and lands that should have been planted for next season's harvest
lay fallow. To the east, in Somalia, competing warlords were quite prepared
to starve their adversaries to death, and foreign military intervention led to
disaster, as Americans well remember.

Government failure to protect people from hunger is not an exclusively
African phenomenon. Reports of widespread starvation resulting directly
from government policy in North Korea, and severe malnutrition in Afghanistan in the chaotic early 1990s form part of this depressing global picture. But politics and ideology are not the only threats to powerless people.
Government economic policies also play a role. It is not enough to produce
a quantity of food for people to sustain themselves; they must also be able to

afford to buy it. Well-stocked markets most of whose local customers cannot pay for a pound of rice do not reflect an absence of malnutrition.

MOMENTOUS TRANSITION

When the English economist Thomas Malthus in 1798 published a warning that the population in Britain was growing faster than the means of sub-sistence, he predicted that population growth would be checked by hunger within 50 years, leading to the disintegration of the social order. For three decades after sounding the alarm, Malthus faced severe criticism from those who saw the future differently, but he gave as good as he got. The exchange is one of the most interesting debates ever recorded. In the end, both Mal-thus and his critics were proven wrong. Food production has not, as Malthus prophesied, increased in linear fashion: it has grown exponentially, recently during the dramatic Green Revolution and today through genetic engineer-ing of an even more invasive kind. And populations did stabilize, though not because of lack of food.

Twentieth-century apostles of Malthusian credo, called neo-Malthusians, forecast scary scenarios of a twenty-first-century planet with tens of billions of inhabitants in a continuous, mortal struggle for the means of survival. The notion of "doubling times" for growing populations was posited as though it conformed with reality: since Brazil in 1970 had about 100 million inhabitants and was growing at 2.8 percent, and since a population grow-ing at 2.8 percent would double in 26 years, Brazil would have 200 million inhabitants in 1996, 400 in 2022, and 800 around the middle of this century. The fact that the growth rates of many European populations were already declining was not deemed relevant: the European demographic "model" was not applicable to the rest of the world. Today, you don't hear much anymore about doubling times. Growth rates virtually everywhere are declining, and no population is likely to be double its present size in a couple of decades.

This is not to suggest that population growth is no longer a concern. The Earth's population overall is still growing rapidly if unevenly, so that the present decade will see an additional 600 million people, most of them in the poorer parts of the world. But it is also clear that the world is in a mo-mentous demographic transition. Today, optimistic projections suggest that the human population will never exceed 10 billion. That is far too many; the neo-Malthusians say that the planet should support no more than 2.5 billion. On the other hand it is far fewer than neo-Malthusian projections of the twentieth century predicted.

The question is how much of the natural world still remaining will survive this transition. Even as Brazil's rate of population growth has declined (it was 1.2 percent in 2004) and its level of urbanization increased (65 percent) the rate of destruction of the Amazonian rainforest has accelerated. Even as African populations ceased growing rapidly as a result of AIDS as well as lower fertility, the rate of deforestation has increased. Before we blame Brazilians and Africans alone for this, remember that the products yielded by the forests, from lumber to leather, are in demand at high prices in the urbanized rich world. In this era of globalization, the consumer demands of wealthy urbanites thousands of miles away can determine the fate of a patch of forest in the tropics.

So this final surge of the population explosion will have key consequences that must be mitigated. But there is now some hope for the future. It is always possible that the whole process will be derailed by some catastrophic event—a global natural disaster such as an asteroid impact, abrupt climate change, a pandemic of some unstoppable disease, an outbreak of nuclear war—but if things continue as they are now, it is conceivable that the world may be planning for a ZPG environment in a half century or so. Of course this will produce its own problems. Who will do the work? How will the costs of longevity and lengthy retirements be paid? What will happen to the concept of Social Security? So far, the European model certainly is no guide. Still, it is conceivable that hunger will at last have been defeated and the gap between the rich and the poor (there will always be one) will narrow.

Meanwhile the world's populations may be modernizing and advanced economies may be globalizing, but old habits die hard. Cultural preferences and religious differences continue to divide a world in which hundreds of millions of farmers still subsist in ways as old as history itself. And despite what you hear and read about the flood of migrants invading the United States, Europe, and other parts of the richer world, consider this: in the middle of the present decade, with the global population numbering 6.7 billion, only about 150 million, or 2.2 percent of the total, live outside their region of birth. Every year, some 2.5 million migrants leave the poorer world and immigrate to the richer countries. The United States is the largest recipient of this stream of migrants: about 1.2 million annually. The next ranking country, Germany, receives about 210,000. UN projections suggest that China, Mexico, India, the Philippines, and Indonesia will be the largest source countries for this global migration network in the decades ahead, but no major increase in the overall volume is anticipated. Indeed, some indicators suggest that the coming slowdown in world population growth coupled with increasing urbanization in the poorer countries will cause a decline in the number of interregional migrants.

Even as we catch a glimpse of light at the end of the population tunnel, another concern arises. After the half century of comparative stability during the two-superpower Cold War, and despite the proxy conflicts that Cold War entailed, the world is now a far more volatile place, in which further nuclear proliferation threatens, civilizational compulsions drive new forms of conflict, and economic globalization meets cultural mobilization on a new and dangerous battleground.

6

THE MESH OF CIVILIZATIONS

The surface of planet Earth may still comprise vast and virtually uninhabited regions where the natural landscape is little changed by human activity and where emptiness defines the cultural atmosphere, but every square mile of these regions lies under the jurisdiction of some government. Canadian authority reaches into the remotest parts of the Queen Elizabeth Islands in the Arctic. China's communist regime asserts itself over the frozen wastes of highland Xizang (Tibet). Chad and Libya carefully monitor their joint border in the desolate heart of the Sahara. The days of unclaimed frontiers and open spaces are over, except on the high seas and in Antarctica. The world today is a jigsaw of nearly 200 countries, independent states ranging in size from vast Russia to tiny Monaco and in population from billionaire China to minuscule Liechtenstein. A few small dependencies survive, leftovers from the colonial era, and some Pacific islands are being argued over by nearby states, but otherwise the appropriation of territory is complete.

So when we open an atlas to the political map of the world, we see a familiar mosaic of boundaries, straight (such as the boundaries that cross the Sahara) and wriggly, for example the jigsawlike layout that defines European countries. You can imagine how this happened: every inch of European territory belonged to some count, baron, or other potentate and was fought over at some point in European history, and the wriggles bear witness to this legacy. But when the colonial powers divvied up the Sahara, what was the point of twisting and bending borders across a vast, mostly empty and barren wasteland? To facilitate acquisition, they drew their boundaries point-to-point, often along parallels and meridians, and not just across deserts, as witness the United States-Canadian border west of the Great Lakes.

Thus we see a framework of national boundaries with which we became familiar from elementary school days onward (at least we geographers hope so). But every now and then it changes in some areas, for example in Europe when East Germany became part of the Soviet Empire and then again when the Soviet Union collapsed and Germany was reunited. Suddenly we had to get familiar with a new layout of Central Asia and with the names (and spellings) of new states: Kyrgyzstan, Tajikistan, Kazakhstan, Uzbekistan, Turkmenistan. And Yugoslavia's disintegration made countries out of formerly obscure "republics"—Croatia, Bosnia, Serbia-Montenegro, Macedonia. Quickly now: does everyone know the difference between Slovenia, Slavonia, and Slovakia?

GEOPOLITICS AND THE STATE

The collapse of the Soviet Union, the breakup of Yugoslavia, the rise of the European Union, the growing strength of regions within states (such as Quebec in Canada), and the emergence of a growing number of multinational economic and other blocs, such as the North American Free Trade Association (NAFTA), have led some commentators to suggest that the days of the state are numbered, that another kind of political-geographical arrangement is in the offing. Don't count on it.

The state will continue to play a pivotal role in international relations for a very long time to come. It is undoubtedly true that provinces, regions, and other subunits play ever-stronger roles in the national affairs of some countries, to the point of influencing policy. This is especially evident in Europe, where regional economic powerhouses such as Catalunia in Spain and Lombardy in Italy are driving forces in Madrid and Rome respectively. It is also true that states are cooperating in multinational organizations to an unprecedented degree, notably in the vanguard European Union but also in other, less ambitious associations. Nevertheless, when national interests are threatened or asserted, it is the state, not some subunit or multinational bloc, that stands its ground in the international arena.

Just consider some recent evidence. The international community may be eager to invoke the nuclear nonproliferation treaty to prevent the further spread of nuclear weapons, but current problems relating to nuclear ambitions (seen as "needs" by the assertive states and as "aspirations" by the rest) come from national governments, not international alliances. It is no coincidence that the three states whose real or suspected nuclear capabilities are causing current concern—North Korea, Iran, and Israel—are all driven by powerful ideologies, but nationalism is what underlies their actions. North Korea's communist and ex-communist allies seek to blunt its nuclear strategy. Iran's European trade partners understandably want to constrain its nuclear program, but Tehran expects no help from its Islamic neighbors. And Israel's nuclear capability is neither acknowledged in Jerusalem nor publicly discussed in Washington.

North Korea, Iran, and Israel happen to be states that aspire to a state model that has long gone out of fashion, the nation-state. North Korea recognizes the historic unity of a Korea that far exceeds its part of the Korean Peninsula, but communist ideology has defined a North Korean nation as well as state and deeply divided it from the South. Iran, as the heartland of Shia Islam, is a state, nation, and religion fortified within a boundary that symbolizes its uniqueness amid adversaries in a turbulent region. Israel seeks to sustain its character as a Jewish nation-state by ensuring an overwhelming Jewish majority through immigration policies.

Although Iran as well as Israel have substantial minorities within their borders—Kurds in Iran, Arabs in Israel—they, as well as North Korea, aspire to a state that exists for the majority. The concept of the nation-state arose in Europe during the period from about 1650 to 1850, when colonial-mercantilist successes coupled with three great revolutions—in agriculture, industry, and politics—created conditions favoring what Europeans called "national states," or countries in which nation and territory were essentially coterminous and where representative government promoted the interest of country and (constitutional) king. States, according to this model, were the territorial domains of more or less homogeneous nations: Spain, Portugal, Italy, the Netherlands, Sweden, Norway, and Denmark were nation-states. The Germans aspired to it; France, the prototype by most definitions of the term, achieved it by doing away with royalty altogether. Some European countries, such as Belgium and Switzerland, never qualified but became reasonably stable states nevertheless. And when, after the First World War, Europeans redrew the map of Eastern Europe, they did not use the model of the West but created discordant ethnic and cultural unions that were destined to break apart, setting the stage for the terrible episodes of ethnic cleansing that attended the breakup of Yugoslavia in the last decade of the twentieth century.

Nation-states, therefore, emerged under particular and unusual circumstances in a corner of the world that was to influence the rest of the globe like no other. Nineteenth-century nation-states tended to be small but powerful, their leaders quite certain that they had achieved the best form of governance. On the wings of colonialism, European concepts of statehood spread far and wide.

By the time that happened, during the eighteenth and nineteenth centuries, European philosophies about the state had already begun to diverge. Most of Europe's nation-states were highly centralized, with a powerful government; these *unitary* states, some still ruled by royalty, included the United Kingdom, Spain, Germany, the Netherlands, and the Scandinavian countries. But from colonial European communities came notions of re-gional-cooperative government, in which some of the power of the center was assigned to the provinces. The notion of federalism seemed quite com-patible with the nation-state model.

Whether unitary or federal, the European models of the state had this in common: they possessed a defined portion of the Earth's surface and were therefore separated by boundaries (not frontiers) from their neighbors; they contained one or more of Europe's nations; they had a governmental system centered in a capital city; and they were served by infrastructures laid out to promote economic development and national power. In an age of colonial competition abroad and intermittent war at home, power was a key ingredi-ent of the successful state.

A German political geographer, Friedrich Ratzel (1844–1904), likened the evolving European nation-state to a biological organism. He argued that the nation, being an aggregate of living beings, would mirror the life cycle of an individual: it would have to be nourished by absorbing other cultures and by expanding into other lands. Colonial acquisition, he suggested, was good for states; defined and delimited boundaries were bad for them. States should be able to vie for territory; when that option was closed, the state, like an old person, would become senile, and the nation would wither. He based this or-ganic theory of the state on historical analysis, and he published his opinions in scholarly journals. But it was not long before these ideas found their way, via his students, into German politics. Later, they became part of Nazi ideol-ogy, promoted especially by a strategist named Karl Haushofer. His brand of *geopolitik* gave that term such a bad name that it went into scholarly disuse for several decades.

In the meantime the European powers were playing power politics on the world stage, laying out, in large measure, the boundary framework with which the world has been saddled ever since. The colonists drew borders where there had been none, built capitals where none had existed, and constructed facili-ties to enable them to exploit and administer their overseas domains. It was, in effect, the first wave of a globalization process that continues, in a different guise, today. Little did they imagine, in the nineteenth century, that the lands they were organizing would someday be on the list of nations of the world.

Nor were colonial domains laid out with the interests of indigenous peoples in mind. Take the case of Africa. As the colonial map of Africa got more crowded, the colonial powers decided to meet and divide the continent rather than fight over it. In 1884, they got together in Berlin to negotiate, and out of that infamous Berlin Conference came, essentially, the map of Africa as we know it today. Only Ethiopia (then known as Abyssinia) and Liberia survived the power grab—Ethiopia because of its natural mountain-fortress protection and Liberia because of its American connections. The rest of Africa was assigned to the British, French, Portuguese, Belgians, and Germans (Spain got a tiny corner, too). The colonial boundaries were drawn, in part, to place African peoples under a single colonial master. But often, the new boundaries divided African nations, splitting them between French and British domains, or German and British, or Belgian and Portuguese.

Next, the colonial powers partitioned their empires internally for administrative purposes, establishing subunits under various names such as "overseas territories," protectorates (surely one of the great misnomers of the time), provinces, and, in the case of the Soviet heirs to the Russian colonial empire, republics. It was inconceivable that any of these subunits might become independent states, so these boundaries too were laid out without much regard for the cultural mosaic. In this regard the French were the great dividers, in Africa as well as in Southeast Asia, their empire eventually fracturing into numerous states with obscure names such as Mali, Upper Volta (now Burkina Faso), Chad, and Laos. The British on the other hand kept populous Nigeria whole and ruled their vast South Asian empire essentially as a single entity, so that India, even following the secession of Pakistan upon independence, did not split into 14 or more independent entities.

Dividing up the world was one thing; transferring the rudiments of what the Europeans viewed as model government was quite another. In the first place, the multicultural landscape of Africa and Asia defied the neat nation-state order the Europeans envisaged. And secondly, the process began much too late, often proceeded under the duress of rebellion and physical obstruction, and was no more popular than American efforts to bring democracy to Iraq. When Britain's Gold Coast approached independence during the 1950s, the scholarly literature was full of analyses of the pending transfer of the "Westminster model" to Ghana, which was to be the country's new name. The naiveté of much of this scholarly discourse was astonishing, right up there with the notion of flower-bearing crowds welcoming United States troops in the streets of Basra in 2003. Of course the model was not transferable in Ghana's case, and the Westminster system soon failed there. It was only the first failure in a long series that brought rapacious rulers to power from Liberia to Lesotho and from Iran to Indonesia.

But whatever the fate of representative government in much of the ex-colonial world, the map was changed forever, laying over the cultural landscape a web of boundaries that spelled power and fortune for some, oppression and disaster for others. The world soon forgets, but the horrific excesses of Amin, Bokassa, Mobutu, Saddam, and Pol Pot—and many others—were made possible in the first instance by the opportunities their bounded territories provided. Taking control in Kampala, Kinshasa, Baghdad, or Pnom Penh meant having power over an internationally recognized country whose very borders constrained intervention from the outside. The world knew that what Verwoerd, Ne Win, Sukarno, Kim, Mengistu and Abacha were doing was wrong, but states are sovereign and so long as what went on remained within their countries' borders, they could operate at will. During the Cold War, the Soviets protected allied dictators at the United Nations; Western interests traded for commodities with countries ruled by onerous despots. No one called it that at the time, but it truly was the New World Order. Very little was heard from the losers, from the Kurds to the Karen.

THE BURDEN OF BOUNDARIES

During the 1991 Gulf War, the inviolability of international boundaries was a big issue. Iraq had invaded and occupied Kuwait with the intent of annexing it as its nineteenth province. Kuwait, and its United Nations allies, insisted that the border between Iraq and the sheikdom was internationally recognized and inviolable. Generals, journalists, and commentators often referred to the "line in the sand" that represented the limit of national jurisdiction, beyond which Iraq should not be allowed to go and behind which it would have to be pushed back.

Boundaries, however, are not lines. That may be surprising, given the maps we see in atlases and on computer screens. There they appear as lines because atlases and screens are two-dimensional. But boundaries do more than partition the surface: they also divide what's below the ground and even the space above. It is best to envision a boundary as a vertical plane, a kind of curtain that hangs above as well as below the ground between two states, dividing airspace, surface, and resources below. Where that curtain intersects the surface, it forms a line. But that line is only one manifestation of the principle.

To illustrate this, Dutch geographers like to tell the story of the coal miners in the southern province of Maastricht, a narrow corridor of Dutch land between Germany and Belgium. Underneath lie rich coal seams that extend from Belgium through Limburg into Germany. The Dutch mined these coal reserves vigorously. They got down to the seam, then excavated it eastward. It wasn't long before they were mining coal below Germany's surface, but of course there was nothing down there to tell the miners that they were crossing the German border. And the Germans had no way of stopping the Dutch underground; they were mining the same seams, but elsewhere. Soon there was some industrial espionage, the Germans got hold of a map of the layout of the Dutch coal mines and claimed violations of their boundary. A pretty good quarrel followed, and the Dutch agreed to stop mining eastward—but only after they had secured a substantial piece of the German reserve.

Things can get even more complicated when it comes to oil and natural gas. Many oil fields and gas reserves extend from one country to another beneath their common boundary. So it was, in fact, where Iraq and Kuwait meet: the Rumaylah Oil Field extends from northern Kuwait into southern Iraq. Here's the problem: when one of the neighboring countries starts to draw on such a joint reserve, oil or gas will flow from the subsurface of the nonexploiter to that of the exploiter. The Iraqis accused the Kuwaitis of drilling oblique wells, so the pipes standing in Kuwait actually fed on oil below Iraq. The charge was never proven. But without doubt, oil flowed below the surface from Iraq toward Kuwait, which drilled much more vigorously into the Rumaylah Oil Field. After the Gulf War, the United Nations redrew the boundary between Kuwait and Iraq, putting all of the Rumaylah reserve within Kuwait and preventing, so it hoped, a future conflict over this reserve.

The Germans and the Dutch also were at it again over natural gas. A substantial gas reserve straddles the border between the Dutch northern province of Drente and adjacent Germany. Massive Dutch exploitation of this underground balloon of gas led to German complaints—but how could the loss to Germany be quantified? Contentious boundary issues, obviously, are not confined to the surface.

Why should a state insist on sovereignty above the ground? There is more here than meets the eye. The now-familiar term "airspace" defines the area above a state's territory where the rights of other states may be limited. For example, commercial airplanes from a particular country can overfly other countries only by official, often reciprocal, agreement. For all the disharmony currently prevailing between Washington and Havana, Cuba allows United States airliners to cross the island on their way to and from Middle and South America. In 1983, a South Korean airliner flying from Anchorage, Alaska to Seoul strayed into Soviet airspace and was shot down by a jet fighter with the loss of 269 lives. There was no retaliation: the Soviets had exercised their right to protect airspace to which they had not given official access.

The international boundary curtain above the ground produces other issues. Imagine this: country A develops a major industrial complex. Clusters of factories consume coal, iron, and other raw materials and produce large quantities of products ranging from toys to heavy equipment. Not much care is taken to protect the natural environment, and a forest of smokestacks belches chemical pollutants into the air.

A large part of neighboring country B lies in the path of prevailing winds, and more often than not, heavily polluted air from country A wafts over country B. When it rains, the precipitation becomes a shower of acid, and plants, forests, water supplies, and other assets of country B are afflicted. Country B objects to this cross-border diffusion of pollution, but the cost of doing anything about it is unaffordable to country A. Country A argues that the atmosphere, like the high seas, is free for all to use and cannot be compartmentalized by boundary planes. Still, country B may press for damages. The stage is then set for still another boundary-related dispute.

Such a scenario is not imaginary. Canada has complained vigorously to the United States about acid rain caused by pollutants emanating from the United States manufacturing belt; Scandinavia's forests and lakes have been affected by acid rain resulting from industrial pollution in Britain and mainland Western Europe. Such events occur with increasing frequency (and are monitored more accurately) than in the past. There is more than one way to violate a boundary!

As if it is not enough to divide what lies below and above the ground on land, states now also have international boundaries at sea. The last frontier of this planet, the world's ocean, is finally being partitioned as the land was. Maritime boundaries, in some parts of the world, are every bit as sensitive and can be as intensely disputed as land boundaries.

It is not difficult to see why. Offshore lie valuable resources, notably oil in many areas of the world. States do not want to see foreign companies drilling for oil or gas just off their beaches, nor do they want foreign fishing boats trawling near the coast. In fact, the notion of a territorial sea is centuries old, but after the Second World War individual countries' maritime claims (prominently including the United States') expanded enormously. This led the UN to organize a series of conferences that produced a Convention on Law of the Sea, stabilizing the situation and creating some order from a chaotic situation. The United States is one of the few nonsignatories to the Convention, but routinely adheres to its terms.

These terms give all states a 12-mile-wide territorial sea (measured in nautical miles), plus a 188-mile-wide Exclusive Economic Zone, where states have certain commercial rights but all ships have the right of innocent passage. In effect, states have a maritime boundary 12 nautical miles offshore; within this boundary all of its sovereign rights prevail. Then they have a second boundary 200 nautical miles offshore, where their EEZ ends. Within this boundary, they can exploit resources or sell exploration rights, harvest, conserve, and manage fish stocks, and to some extent control pollution, smuggling, and other activities, including those that might affect national security. Beyond the EEZ boundary lie what remains of the high seas, the open ocean not owned or controlled by any country or countries. The Convention does include a clause asserting that resources in this maritime frontier should be regarded as the "common heritage of mankind," which is one reason the United States has declined to sign it, but that language has been mitigated and considerably qualified in later revisions.

From these details one concern is immediately evident: if states can claim a 12-mile territorial sea, numerous straits and narrows, from the Dardanelles in Turkey to the Bab-al-Mandab at the southern Red Sea entrance, are wholly owned by coastal states. And whole seas, from the Caribbean to the North Sea, are covered by adjoining EEZs. And that is exactly the situation today. While ships of all flags are supposed to have freedom of navigation through straits narrower than 24 nautical miles, the fact is that they are traversing sovereign seas and can be subject to inspection under certain circumstances. While ships can cross EEZs at will, states can restrict their routes, for example in areas where oil rigs rise above the water (the North Sea is a veritable forest of such installations).

So the state now projects itself onto the seas, in places even marking the limits of its maritime jurisdiction by means of buoys. Maritime boundaries can be difficult to define, because they are measured from irregular coastlines where tidal ranges affect the points of calculation. These measurements can mean a great deal when it comes to the rights of states heavily dependent on fishing, and normally friendly states sometimes quarrel bitterly over a few hundred yards of maritime domain.

Boundaries, then, must serve multiple purposes ranging from immigration regulation to resource allocation and from airspace control to maritime security protection. They are a reality of life for anyone involved in international travel or business. We all know how divisive and disruptive they can be, but they are here to stay, even where serious attempts are under way to reduce their impact through multinational agreements and treaties. The question is, if good fences make good neighbors, do good boundaries make for good relationships among states?

BOUNDARIES AND THE STATE SYSTEM

Political geographers think of the state as an open system consisting of numerous interacting parts, with inputs and outputs involving migration, trade, finance, transportation, information, and many other functions. Some of these parts are visible on an ordinary map: the capital city, the so-called core area where population density tends to be high and where most of the country's productive capacity is concentrated, transport networks that look like the arteries of a complex organism, and external boundaries that mark the limit of jurisdiction (but not necessarily influence).

We humans tend to be quite possessive about our territories. Many years ago Robert Ardrey published a popular and much-debated book in which he argued that we have an innate need to divide and subdivide our activity space, calling it our "territorial imperative," encoded in our genes and shared with many animals (Ardrey, 1966). In my experience, this imperative varies culturally, and thus regionally. In the Netherlands, where I grew up, every inch (centimeter, there) of land seems to be walled, fenced, or hedged off. When I arrived in Evanston, Illinois via several years in Africa, I was amazed to see houses separated by lawns and flower beds, but without any enclosures or gates, even along the sidewalk. Make no mistake, Americans can be very territorial, but American cultural landscapes suggest that this is less of an obsession in spacious America than in the crowded Netherlands. Population density probably has something to do with it, as well as level of income, but whatever the reason, spatial variation does mark our territorial possessiveness. I have worked in rural areas of Africa where fencing off farmland is simply not done.

When it comes to the national territory, however, even those people living on fenceless properties can become emotional, even irrational about borders. Governments sometimes have boundary disputes, but they seem to be able to act reasonably toward their neighbors only at their own peril. I happened to be in Russia in the mid-1990s when then-president Boris Yeltsin suggested that it was time to negotiate with the Japanese over a group of islands occupied by the Soviets ever since the end of World War II and claimed by Tokyo, the so-called Northern Territories at the western end of the Kuriles off Hokkaido. These islands were not part of Russia before the war; the Russians simply never left afterward, as a result of which Russia, heir to the Soviet Union, and Japan were still technically at war a half century later. The Japanese had proposed a massive development scheme for Russia's Far East in return for Russia's withdrawal from the islands, and Yeltsin thought it was a good idea and said so in the Duma. The next day the English-language Moscow paper thundered "NOT ONE METER" and the Russian-language press used even stronger language. Not one square meter of "Russian territory" would be yielded to the "blackmailers" in Tokyo. So much for reason and sanity. And feelings haven't changed. In late August 2004, Japanese president Koizumi sailed past the islands, just outside the 12-mile limit, and the next day the Russian press accused him of "aggressive behavior toward the motherland."

Technically this is a territorial dispute, not a boundary dispute, but boundary issues can evoke similar emotional reactions, and worse. When you see the hardscrabble countryside where in 1998 and 1999 thousands of Ethiopian and Eritrean soldiers lost their lives over a few square miles of land and some isolated villages it is enough to make you weep. But the world is full of actual and potential trouble spots like this. While many thousands of miles of international boundaries are carefully defined, other segments are not, and where a boundary is not well buttressed by legal definitions, trouble lurks. Cooperative neighbors turn into fierce enemies when they disagree over joint borders, as happened between Ethiopia and Eritrea—whose leaders were friends before the crisis erupted.

Establishing a well-defined boundary is a bit like preparing a property survey during a real estate transaction. Anyone who has bought or sold real estate has had this experience. I have always been impressed by the detail in the resulting documents: my waterfront house in Coral Gables, Florida, required a full page of legal definition plus a large, detailed map. Next I bought a property in Georgetown (Washington, D.C.). It was just a small row house with a tiny garden in the back. Two surveyors spent most of a day working up the survey and description, which, weeks later, was part of the closing documents required to complete the purchase. Then, less than a year later, I refinanced the house. Believe it or not, another survey was done. The house hadn't moved, no construction had occurred, everything was as it had been. But the property was mapped once more.

If a tiny property such as mine could require such meticulous attention, imagine what is needed to make legal the surface boundary between two countries! That's exactly what those surveyors were doing: they prepared the legal description of the bounds of my property. For an international boundary to have validity, it needs definition in legal terms. This means that it must, in the first instance, be described in terms of the physical and cultural features that mark or affect it. Before a mapmaker can draw a single line, before governments can drive a single post into the ground, the boundary's legal definition must be in place.

So this is the first, and only the first, stage in the establishment of a good-fences boundary between states: its *definition* in legal language. If you want an insight into this process, it is worth looking up in the library a two-volume work by Edward Hertslet titled *The Map of Africa by Treaty* (Hertslet, 1909). He tells the story of Africa's partition and quotes numerous boundary-treaty definitions. A sample might read like this: ". . . from the fencepost on the southeast corner of the property owned by William Morris Davis approximately due eastward along the crest of the ridge locally known as Bowman's Rise 5783 feet (1792 meters) to the center of the watergap occupied by the Semple River. The boundary follows the center-line of the Semple River southward (downstream) approximately 6.13 miles (9.9 kilometers) to its confluence with the Ratzel Creek at which point the boundary becomes the west bank of the river. Precisely 695 feet (211.836 meters) from this point the boundary is marked by the center of the large Baobab tree from where it turns 135 degrees in a straight line for 3.47 miles (5.58 kilometers) . . ."

Next comes the job of putting the legal language and the surveyor's work on the map. This is referred to in legalese as *delimitation*, and it may oc-cur years after the two neighbors agreed on the definition. The associated problems are obvious: in the example just cited, that baobab tree may have been knocked over by elephants, erosion may have shifted the streams. If the boundary area is sensitive, it may be necessary for cartographers and survey-ors to go back to it to determine the intent of the original language. When that happens, get ready: interpreting intent can be a ticket to trouble.

There's still another, common problem. Governments of countries that were once colonies may be unwilling to accept the treaty definitions writ-ten by the colonial powers. Hertslet's book on African boundaries is basi-cally the story of colonial acquisition complete with land grabs and ruthless exchanges. Africans had no say in the process, and while modern African governments generally have accepted their inherited boundary framework, some disputes have erupted, notably in resource-rich areas. Nigeria and Cameroon clashed over their joint boundary in the Bakassi Peninsula, where oil fields are at issue, and it took a decision by the World Court in The Hague (in favor of Cameroon) to settle the matter.

In other instances, colonial-era boundary treaties cause trouble because a present-day government argues that its predecessor, although perhaps con-sulted, was coerced into agreeing to it. An important case in point is the boundary between India and China in Xizang (Tibet), resulting from the Simla Convention dating from 1914. The powerful British wanted to recog-nize the McMahon Line, running along the main crestlines of the Himalayas in the area east of Bhutan, and they convened a conference in the town of Simla to draw up the treaty. Although Chinese representatives at the meeting approved it, the treaty was rejected when the draft reached China's capital. The British simply overruled that rejection and only they and the Tibetan representatives signed it, making Arunachal Pradesh part of their South Asian empire. But the Chinese did not forget, and in 1958 the new com-munist government began a campaign to reposition the boundary where its Simla delegation had wanted it. The issue remains unresolved to this day.

The McMahon Line, needless to say, does not appear on official Chinese atlases or textbooks.

Disagreements over boundary definitions involve not only China and India but also China and Russia, China and Kazakhstan, India and Pakistan, and numerous smaller countries. The terrible ten-year war between Iraq and Iran began not as a religious conflict, as many suppose, but as a boundary conflict in the area of the Shatt-al-Arab. Even as boundary issues are settled by negotiation, for example between Peru and Ecuador during the 1990s and Nigeria and Cameroon more recently, others arise, some involving maritime boundaries. With the global web of borders has come a worldwide source of conflict.

A final stage in the establishment of boundaries is sometimes deemed necessary for purposes of immigration control, security, or other perceived needs of the state. This involves marking the boundary on the ground in some way, to ensure that it is observable, to control movement, or even to defend the national territory. This process, the *demarcation* of the boundary, affects a small percentage of the global framework, but it is usually contentious where it does.

Extreme examples of demarcated boundaries past and present include the Berlin Wall, the DMZ between North and South Korea, the fence along stretches of the United States–Mexican border, and the "Security Wall" between Israel and the West Bank. More often, demarcation is accomplished by concrete markers, posts, or other signals that alert travelers that they are crossing an international boundary (the informal rule is that any such marker must be visible on the ground from both of its neighbors, so that any transgressor cannot claim to have wandered across the border inadvertently). Usually, demarcation to inhibit movement reflects a crisis situation, as is the case with cross-border Mexican immigration to the United States and terrorist infiltration in Israel, and it tends to signal problematic relations between neighbors.

BOUNDARY HIERARCHY

The world map of international boundaries reveals legality and conceals reality. Everyone who travels internationally knows that some borders are crossed with ease, others with great difficulty. Still others cannot be crossed at all. Commuters cross the border between Canada and the United States, but not the border between the United States and Mexico. You can drive from Barcelona to Berlin without slowing down for customs and immigration, but don't try driving at speed across the boundary between Poland and

Belarus. It is a lot easier to cross the border between Singapore and Malaysia than between China and Vietnam. This phenomenon repeats when you travel by air. Kuala Lumpur to Bangkok is about as routine as Chicago to Toronto, but Shanghai to Moscow (actually, pretty much anyplace to Moscow) can be, no pun intended, a bear.

When I draw a map of the "easy" and "difficult" boundaries with which I have some personal experience, augmented by the observations of colleagues who do fieldwork in foreign areas, an interesting pattern emerges. The world seems to be divided into about a dozen realms within which boundaries are usually, though not always, reasonably "easy," but between which they tend to be tough to cross, surface or otherwise. These geographic realms match the way we think intuitively about the layout of the modern world when we refer to such realms as East Asia, North America, Europe, or Subsaharan Africa (Fig. 6-1). The world may count upward of 200 countries today, but the web of borders reveals regional clusters of states that share cultural histories and a global set of geographic realms that have much less in common.

It is tempting to conclude that this represents a prospective simplification of the global boundary framework, that intrarealm boundaries are likely to fade even as interrealm boundaries lose their divisive functions, but such a conclusion would be premature. Europe's regional integration is unique and unmatched, even in two-state North America. Elsewhere in the world, shared cultural histories and attributes have not generated anything similar. Just what the implications of Figure 6-1 may be remains uncertain.

GLOBAL CIVILIZATIONS—MESH OR CLASH?

The prospect of civilizational clashes as the ultimate form of conflict that has taken us from clan warfare and tribal strife through feudal contests and interstate hostilities to two world wars has been a tenet of political geography for decades. Figure 6-1 appeared in introductory geography textbooks as early as the 1970s and has been a standard topic in political geography for even longer (de Blij, 1971). Geographers in the 1980s forecast the Soviet Union's loss of control in Eastern Europe and the collapse of the U.S.S.R. itself, but not until the essence of Figure 6-1 appeared in Samuel Huntington's *Clash of Civilizations* did the intelligentsia pay attention (Huntington, 1996). The argument that world politics is being reconfigured along cultural fronts, and that civilizational conflict will erupt as cultural "fault line wars" (a rather unfortunate allusion to geomorphology inasmuch as a fault line is an erosional remnant rather than the real thing) was not new in the 1990s, at least not in geography. But that merely serves to highlight Huntington's

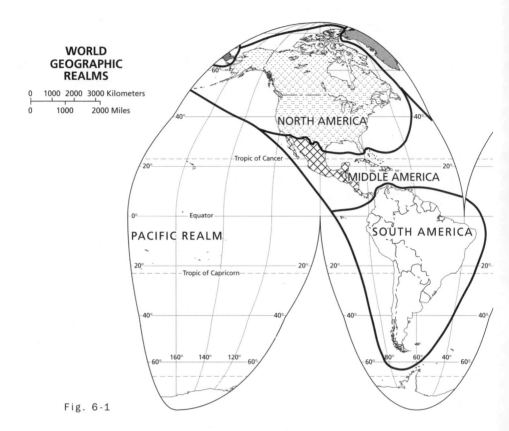

Fig. 6-1

thesis, which holds that the remaking of the world order in the twenty-first century will essentially pit the "West Against the Rest." The realms of intuitive geographic consensus of the 1960s have become far more clearly bounded, yielding culturally discrete entities whose internal cohesion is in large part based on growing antipathy to the West. The West of the 1960s, meanwhile, has become quadripolar, with an assertive superpower United States in an increasingly contentious relationship with a wary Europe and tendentious ties to unpredictable Russia even as Brazil and greater South America loom larger on the strategic stage.

The map suggests something else: if a dislike of the West is uniting countries and cultures in other geographic realms, then growing threats to the multipolar West are bound to have a similar effect in the West itself. Until September 11, 2001, such threats had relatively limited impact in the Western Hemisphere (the largest Islamic terrorist strike was in Argentina in 1994), and even after 9/11 Western countries could afford the luxury of disagreement over jurisdictional and legal issues involving alleged terrorists and over extralegal incarcerations and military tribunals. Those niceties are likely to

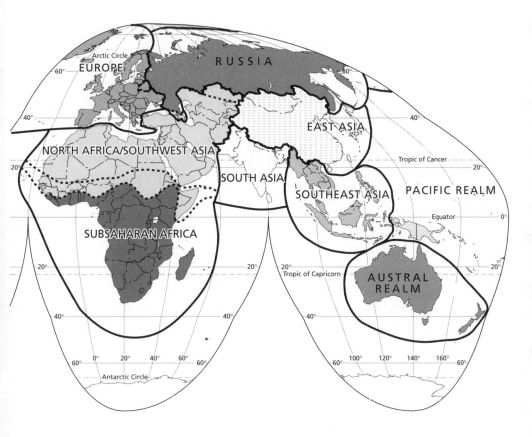

disappear when the phrase "war on terrorism" becomes a matter of necessity rather than choice.

In any case, the Islamic realm as mapped in Figure 6-1 is not a civilizational entity engaged as such in this "war," nor are any of its state governments overtly participating in it (some, such as Libya, have committed terrorist acts and others, like Iran, have supported terrorist organizations). It is true, as Huntington argues, that the boundaries of the Islamic realm, external as well as internal, are especially conflict-prone "fault lines," but as a geographic entity this realm shows no prospect in a ranking of the world order. One Islamic government now possesses nuclear weapons and others, probably beginning with Iran, may acquire them, and the possibility of a surprise attack on an Israeli or Western target or ally at some future time is not inconceivable, but a coordinated "civilizational" campaign is not a prospect.

The real short-term and long-term challenges to Western supremacy will come from China and India respectively. Only the United States dominates its geographic realm as China and India do, but the United States for all its power has just a fraction of the population of either. There was a time when the Mediterranean Sea carried more trade than any other body of water, and next it was the North Sea, and then the Atlantic. Now the era of the Pacific has arrived, and both the United States and China have major frontages on this greatest of all oceans. Already, trade across the Pacific, in value as well as volume, now exceeds that crossing the Atlantic. And on the Pacific side, China is emerging as the world's most voracious consumer, competing with the United States for resources ranging from oil to ores. With a vast manufacturing base and a hugely favorable trade balance not only with the United States but also with the rest of the world, China will become regionally dominant in Southeast Asia, will confront the United States presence as a stabilizing force in the Western Pacific, and will mount a global challenge to America and the West with the first real prospect of altering the world order.

If what may become a Cold War with China can be kept in check, the greatest long-term beneficiary is likely to be India, whose promise as an economic power, civilizational force, and cultural icon in what will become the most populous geographic realm on Earth is unmatched. India faces short-term challenges not confronted by China, including the integration of its huge Muslim minority, and for all of its democratic credentials its successive governments have failed to steer a sound economic course or to combat costly corruption. But India, like China, is a nuclear power, a major military power overlooking a great ocean, and a rising regional force. It will play a major role in the remaking of the world order.

STATES AND CIVILIZATIONS

Looking again at Figure 6-1 in context of the mesh of civilizations, it is clear that any transformation of the world order during the twenty-first century will still be primarily the function of states, not "civilizations." The putative contestants will be the United States, China, and India. By no stretch of the geographic imagination will politically fractured civilizations such as the European, Islamic, Southeast Asian, or African play comparables roles— and of these four, only the Islamic realm has produced extremists seeking to alter the course of history through targeted and random violence. The world's boundary framework creates four megastates that are approximately coincident with "civilizations," including Orthodox-Christian Russia whose cultural and demographic decline render it an unlikely participant in the process of global transformation except as a flank of the West. Three of these four states will determine the course of events, and it is China's historic durability and Britain's unification of India that have brought us to this momentous prospect.

7

RED STAR RISING:
CHINA'S GEOPOLITICAL GAUNTLET

Napoleon, who knew a thing or two about empires and imperialism, is supposed to have remarked that China was a giant asleep, and that whoever woke it up would regret doing so. In the two centuries that followed, European colonial powers shook China's lethargic rulers, Japanese armies jolted the Chinese heartland, and Soviet ideologues got in bed with their Maoist counterparts. But China outlasted the colonialists, ousted the Japanese, and outdid their Stalinist advisors' communist fervor, all without entering the global stage. Nor did the Chinese retaliate against the Europeans, even allowing the British to reassert themselves in postwar Hong Kong. No state-sponsored violence was directed against Japan, whose dreadful wartime atrocities rouse Chinese passions to this day. And the Soviet advisers were simply sent home, not punished for their despised "revisionism." All the while, China was not a member of the United Nations, Chinese goods were not seen on international markets, and China slumbered on in self-imposed isolation. Convulsive nightmares such as the Great Leap Forward and the Great Proletarian Cultural Revolution, costing tens of millions of lives, periodically stirred China but had no effect on the outside world. Famines and natural disasters had little impact on population numbers: overwhelmingly rural China grew exponentially, contributing about one-quarter of the global yield of the twentieth-century population explosion.

But then, on a fateful day in 1971, President Nixon announced that he had sent an emissary to the Chinese capital to make arrangements for a visit to meet with the country's premier, Zhou Enlai, and with its Communist Party Chairman, Mao Zedong. To most Americans it seemed inconceivable that an American president would sit at a table with the rulers of the world's largest and most virulently communist nation, one whose army had killed many American soldiers during the Korean War just two decades earlier. Yet even before the visit took place, the United States announced that it would support action at the UN General Assembly in favor of seating the People's Republic of China. Although the United States also proclaimed that it would oppose the ouster of Taiwan from the UN, the resolution that admitted China simultaneously stipulated the expulsion of the Taiwanese delegation. Clearly China was about to enter the global stage, with incalculable consequences. When Nixon arrived in Beijing on February 21, 1972, it was not difficult to hear the echo of Napoleon's prescient words.

TECTONIC INTERVENTION

China in the 1970s still was embroiled in the Great Proletarian Cultural Revolution, started by Mao and his clique in 1966. Fearful that Maoist communism had been contaminated by Soviet "deviationism" and worried about his own stature as its revolutionary architect, Mao unleashed a campaign against what he viewed as an emerging elitism in Chinese society. He mobilized young people living in cities and towns into cadres known as Red Guards and ordered them to attack "bourgeois" elements throughout China, called on them to criticize Communist Party officials, and encouraged them to root out opponents of the system. He shut down all of China's schools, persecuted intellectuals deemed untrustworthy, and motivated Red Guards to engage in what he called a renewed "revolutionary experience."

The results were disastrous. Red Guard factions took to fighting among themselves. Anarchy, terror, and paralysis followed. Thousands of China's leading intellectuals perished, moderate leaders were purged, and senior teachers, elderly citizens, and older revolutionaries were tortured to make them confess to crimes they had not committed. As the economy suffered, food production and industrial output declined. Violence and famine killed as many as 30 million people as the Great Proletarian Cultural Revolution spun out of control.

The aftermath of all this hung over China during Nixon's visit, but it was not a topic of conversation. Neither China's nor the United States' internal affairs were discussed as a matter of protocol. After Nixon left, the waning Cultural Revolution continued to afflict the country, and as Communist Party chairman, the powerful Mao could pursue his objectives at will, constrained only by the more moderate Zhou. There is no doubt, however, that the Party's status and reputation suffered as a result of the Cultural Revolution. The arrant behavior toward elders by youngsters in a culture that for millennia had revered its senior citizens aroused widespread resentment and disdain. The resulting dislocation left few citizens unaffected. Confidence in Party leadership was at a low ebb.

What caused the unimaginable, the end of Mao's rule? Historians ascribe it to his failing health and loss of control during his final year, and to the power struggle that shook party and country in the mid-1970s. Undoubtedly the Cultural Revolution had left the state as well as the Communist Party in disarray. But geographers point to another factor that undoubtedly hastened the process. On July 28, 1976, twin earthquakes struck near the city of Tangshan, about 100 miles (160 km) east of Beijing. Tangshan, then a city of about 1 million, was almost completely destroyed. Loss of life and physical destruction also ravaged Beijing's port city, Tianjin, as well as parts of Beijing itself. The death toll, never confirmed by Chinese authorities, may have exceeded 700,000, making this the most deadly natural catastrophe of the twentieth century by far. So disorganized, incompetent, ineffective, and corrupt was the relief effort that word of it spread throughout a China already spent by a decade of Mao's Cultural Revolution. The Chairman for Life died on September 9, just six weeks after Tangshan. It probably was no coincidence.

If Nixon's opening to China caused misgivings in the West (and, to be sure, in the Soviet Union, no longer on China's ideological page), it also contributed to qualms in China itself. Mao adamantly opposed changing the political or the economic system that had made the People's Republic what it was. Zhou was in no position to change Mao's mind, but as premier he was able to promote party leaders who took a more pragmatic view of China's needs. One of those leaders was the man would guide post-Mao China from torpor to tiger: Deng Xiaoping. In the fateful year of 1976, when power struggles and natural disasters buffeted the state, Deng readied policies that would combine continued communist dictatorship with market-driven economics. Within decades, not distant generations, China was a power to be reckoned with on the global stage. By the turn of the century it was a nuclear power, an economic heavyweight, a military juggernaut with the world's largest standing army and a growing navy, and a political force with increasing influence in its own backyard and beyond.

WAS IT INEVITABLE?

Most geographers were not surprised by China's ascent any more than they were by the Soviet Union's disintegration. The position of Eurasian states in global geopolitics had been a topic of debate for more than a century and had generated some penetrating analyses and remarkable forecasts. The first shot in this productive exchange was fired on a snowy night in January 1904 in London, when a Scottish geographer named H. J. (later Sir Halford) Mackinder delivered a lecture titled "The Geographical Pivot of History." The winter weather kept his audience rather small for a major event at the Royal Geographical Society, but from the minutes of the meeting it is clear that those present were aware that they had heard something extraordinary. When the paper was published in the *Geographical Journal* some months later, complete with a transcript of the discussion that followed, it raised a storm of reaction and for decades afterward was the most often cited article in geography (Mackinder, 1904).

Essentially, Mackinder argued that while British naval forces had made Britain the leading superpower of the time, impregnable land-based strongholds would come to dominate geopolitics. A quite exhaustive analysis of resources and risk factors led him to conclude that the Pivot Area, that is, interior Eurasia, contained the wherewithal to allow a future power to successfully challenge for world domination. This Pivot Area—he later called it the Heartland, and his idea became known as the Heartland Theory—lay in Eastern Europe and western Russia. At a time when Russia was a collapsing, overextended country losing a war against Japan on its eastern flank, Mackinder foresaw its emergence as a world power. When the Soviet Union emerged after his death in 1947 as one of the two power cores in a bipolar world, Mackinder's original article got a lot of posthumous attention.

Until the year of his death, Mackinder participated vigorously in the ongoing debate his Heartland Theory had generated. Perhaps Mackinder's most perceptive critic was a geographer named N. J. Spykman who, in the 1940s, published a book in which he compared the assets of Mackinder's Heartland with those of the Eurasian Rimland (Spykman, 1944). He probably was the first observer ever to use the term "rim" for the Eurasian periphery, foreshadowing the commonplace "Pacific Rim" to which we refer today. Spykman concluded that no matter how powerful a Soviet Union would emerge from the Second World War, its strategy should be to extend its domination into the Eurasian Rim: failing this, a Eurasian Rim power or alliance of powers would ultimately and successfully challenge for supremacy.

There is an interesting geographic order to what has happened in Eurasia over the last several thousand years. In the west, the succession of power proceeded on a latitudinal ladder, from Egypt to Crete to Greece to Rome and finally to Western Europe. Ever larger spheres of dominance attended this sequence, from the Egyptians in North Africa to the Greeks of Alexander and Persia to the Romans and their Mediterranean empire and, ultimately, to the Western Europeans and their global colonial realms. But then the succession took on a longitudinal dimension. Insular Britain was the world's first naval superpower. Mainland Germany mounted two major wars, the second for world domination. Next came the Russian (Soviet) challenge, the Heartland power at its zenith. Is it now time for the cycle that began on the western perimeter of Eurasia to reach the eastern rim?

The United States, from its ocean-moated national fortress, crucially in-fluenced the course of events affecting all three of these powers, beginning with its obstruction of British colonialism. The Second World War and its aftermath in Korea left America on the doorstep of East Asia from Japan and Taiwan to the Philippines and Vietnam. The Cold War marked a half century of bipolar geopolitics and, when it ended in the disintegration of the U.S.S.R., left the United States disproportionately powerful in a world where, until the rise of China, there would be no clear Number Two.

POWER ON THE PERIPHERY

China's takeoff during the past 25 years unleashed its economic potential and witnessed the greatest short-term regional transformation in the his-tory of humanity. "Pacific Rim" became bywords for reconstruction, reno-vation, reformation, modernization at a dizzying pace. The whole of coastal China, it seemed, was one vast construction site from Guangzhou to Dalian. Adjoining Hong Kong, the fishing village of Shenzen with its duck ponds and pig pens became a skyscrapered metropolis of 3 million in less than two decades, the fastest-growing city in world history. Across the Huangpu River from Shanghai's historic colonial avenue known as the Bund, the rural countryside was swept away and the futuristic skyline of Pudong re-placed it. Gleaming new airports, high-speed intercity highways, multilane urban ringroads, state-of-the-art suspension bridges, hydroelectric proj-ects, vast factory complexes, and modern container-port facilities reflected China's burgeoning economic power. Foreign investment from Japan, the United States, even (indirectly) from Taiwan propelled the process, and on the insatiable American market Chinese goods sold in quantities unimag-ined before, creating a huge trade surplus for Beijing and financial reserves undreamed of even by Deng Xiaoping himself.

With economic strength comes political clout, and by the turn of the century the outlines of a new geopolitical relationship between the United States and China were visible. After the Second World War, the Japanese had become allies of the United States; after the Korean War, South Korea achieved democracy as well as economic success. American troops remained based in both Japan and South Korea. The United States role in protecting Taiwan from Beijing's designs stopped short of stationing American forces there, but United States warships patrolled the waters between mainland and island during times of tension. The United States had military bases in the Philippines (now vacated) and Singapore, fought a war in Vietnam without significant Chinese involvement, and maintained a western-Pacific presence from Guam to Palau. As Chinese students never tire of telling me when I visit China, the United States is still on China's doorstep, but China is not on America's.

The United States, as the public learned in 2001, also conducts intelligence operations in the western Pacific, as close to China as international law allows (and undoubtedly closer). On April 1, a United States spy plane flying just outside the 12-mile limit collided with a Chinese fighter jet that was tailing it; the fighter plane went down but the spy plane, though damaged, landed on Hainan Island where the crew was briefly detained. The incident aroused much public anger in China, and although it was quickly settled its impact on public opinion in China lasted longer. This and other provocations such as the mistaken bombing of the Chinese embassy in Yugoslavia by United States warplanes during the Kosovo crisis (at fault was an outdated map: they hit the wrong address) all contributed to still another byproduct of economic success and strategic frustration: a rising tide of nationalism. I have watched and experienced this upsurge during 25 years of very nearly annual visits to China. The perceived superpower arrogance of the United States, its omnipresence in the region, its role involving Taiwan, criticism of China's human-rights practices, refuge given to dissidents, and other irritations drive nationalist sentiments expressed in newspaper editorials, letters to editors, public reactions to perceived slights, and in virulently anti-American and nationalistic best-selling books such as *China Can Say No* (Song Qiang, 1999).

I can attest to the rise of Chinese nationalism on several fronts. My geography textbooks are sold worldwide, and have been translated into Chinese as well. I receive a steady stream of critical comments from students and other readers at home and abroad, including many from Chinese students now enrolled at United States universities and colleges, about the way their countries and regions are represented. Chinese readers tend to be especially displeased by my maps: these do not clarify that Taiwan is in fact a province of the People's Republic, do not show parts of India and Kazakhstan to be Chinese territory, do not, as one angry correspondent wrote, reveal Russia's Pacific Southeast as "stolen lands," and do not assign disputed Pacific islands to China. On the other hand, a Manchu mindset seems to be as strong in Taiwan as in Beijing: my carefully nuanced narrative on Tibet (Xizang) evoked a passionate denunciation from Taipei. "The boundaries established during the Qing Dynasty are the irrevocable borders of China, and you have no right to judge our historic heritage."

Will the United States and China find themselves in a geopolitical confrontation in the decades ahead? Certainly the potential exists. When the topic arises in discussion in China, the issue of Taiwan inevitably comes to the fore. Chinese mainlanders overwhelmingly regard this as a domestic matter to be resolved by negotiation if possible but by force if necessary; American protection of Taiwan's quasi-independent status is despised and resented. Under the umbrella of American security guarantees, Taiwan has not only moved from autocracy to democracy but has begun debating the merits of independence, a prospect that provokes Chinese threats of armed intervention at any cost, even military confrontation with the United States. Although President George W. Bush, during a visit by China's leader Hu Jintao in 2002, reiterated American disapproval of any such action by Taiwan, Washington may not be able to constrain Taipei's political initiatives. Meanwhile the United States sells Taiwan modern weapons. Over the issue of Taiwan, a small-scale Cold War is already in progress, and it is fraught with risk. The United States is committed to protect Taiwan against military intervention; the Chinese are placing rockets aimed at Taiwan on the mainland side of the Taiwan Strait. Even as Taiwanese investment helps drive the economic growth of China's Pacific Rim and tentative low-level discussions go on between Taipei and Beijing, political leaders engage in provocations and militaries flex their muscles. It is a recipe for trouble.

Debate this issue in China, and tempers tend to flare (a Chinese student once asked me how Washington would like it if a "bunch of terrorists" took over Puerto Rico and China prevented the United States from blockading and regaining the island). Recently, though, another matter has come to the fore: the implications of George W. Bush's "axis of evil" and the invasion of Iraq. Communist China's communist neighbor and long-term ally North Korea lies at the far end of the "axis" but on Beijing's doorstep, and when regime change for Pyongyang is mentioned in Washington, the notion is not exactly welcome in the Chinese capital. Chinese leaders are obviously troubled by North Korea's efforts to assemble a nuclear-weapons arsenal, but unilateral intervention by the United States would undoubtedly precipitate a political crisis in China. As it is, American determination to export democracy to countries that do not have and may not want it creates unease in this, the world's largest—and still communist-ruled—nation. The memory of the 1989 prodemocracy movement violently quelled on Tiananmen Square may be fading, but it is not expunged.

Competition between the United States and China is not confined to the political arena. Watch Lou Dobbs's almost daily CNN drumbeat on American jobs leaving for foreign places and the economic disadvantage in which the United States finds itself in its trade relations with China, and you would think that the new Cold War is not only here but already lost. Far more consequential, however, will be the unavoidable byproduct of China's economic growth: an escalating demand for energy resources at a time when supplies are unusually limited and less secure. Not many years ago, China was able to meet its energy needs from domestic sources, mainly its oil reserves in the far west and from coal deposits in several parts of the country. Today, China must import oil, and in fast-growing quantities; its demand is having a major impact on global energy trade. The most direct supply lines are from the Caspian Sea area and from Russia, and, fortunately for the Chinese, only one (though territorially huge) country lies between China and the Caspian Sea: Kazakhstan. Agreements between Beijing and Astana have allowed for the start of construction of pipelines from the Tengiz oil reserve in the Caspian Basin to Urümqi in western China. Meanwhile, the Chinese are negotiating with the Russians to extend Russia's pipeline network into China's northeast, but here the Chinese are meeting competition from the Japanese, who view Russia's eastern oil reserves as vital to their future. Most of Japan's enormous demand for oil and natural gas is met by long-range tankers, but disruption of supply lines is a growing risk in this era of political instability and terrorist threats in the overseas source areas. Access to supplies from the comparatively nearby Russian oil fields is crucial to Japan's economic prospects. The Japanese want Russia to provide oil via a coastal terminal across from Japan. However this process plays out, China's emergence as a large and growing consumer of a nonrenewable global resource of which the United States is by far the world's largest consumer and of which America has only limited domestic reserves will affect relations between the two states as time goes on. The question is not whether these relations will become more difficult; rather, it is how to prepare for and manage the difficulties when they arise.

CHINA'S SURPRISING GEOGRAPHY

One way to mitigate the problems ahead is to comprehend them better. In the introductory chapter I argued that America's rather sudden ascent to the status of sole superpower confers on Americans a responsibility to further internationalize their outlook, to be better informed about the countries and societies on which the United States has such an enormous impact—and that one good way to go about this is to study global geography. Surveys indicate that Americans' geographic knowledge of China remains dim, despite our deep involvement in the burgeoning of its economy. In my own experience, though, we tend to be fascinated and often surprised by even the basics of China's regional geography.

Take just one example: the map of the conterminous United States superimposed on a map of China (Fig. 7-1). This map is not only to scale, it

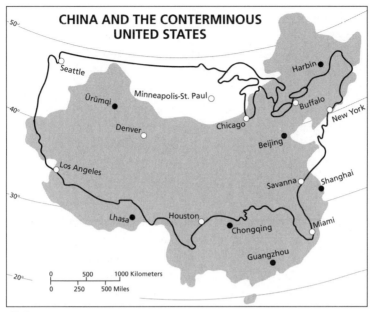

Fig. 7-1

also puts the two countries at their appropriate latitudes. Many Americans think that China is much larger than the United States, given its huge population. But in fact, including Alaska, the United States is slightly larger (3,717, 796 against 3,696,521 square miles). In the east, though, China extends farther north even than the latitude of northern Maine, and farther south than southern Florida. China's northeastern provinces have bitterly cold winters and short summers, and a sliver of Russian territory landlocks them from the Pacific Ocean. In the south, China is truly tropical: the island of Hainan lies in the general latitude of the Greater Antilles. Hong Kong and Guangzhou lie well south of Miami.

Westward, both countries become more mountainous and drier, although more gradually in the United States than in China. In the United States, relatively mild conditions prevail into the Great Plains; in China, forbidding, ice-capped mountains in the south and vast deserts in the north dominate a physical landscape with comparatively few oases of cultural life. Although it is about as far from Shanghai to Ürümqi as it is from Savannah to Los Angeles, there is no Los Angeles in China—nor is there a west coast. China's far west abuts the steppes of Mongolia and the mountains of Turkestan, historic frontiers of contact between Chinese and non-Chinese in remote inner Asia.

Which brings us to what is surely the best-known aspect of China's geography: the country's huge human population, now exceeding 1.3 billion, more than one-fifth of all humanity and, despite a declining growth rate, still adding about 9 million annually. With such numbers, you would imagine that every part of China is densely peopled, but in fact the opposite is true. Ninety-five percent of China's population is concentrated in the eastern one-third of its territory, and even in this eastern third there are vast, sparsely populated areas. Natural landscapes have a lot to do with this—people tend not to agglomerate in icy mountains or in arid deserts—but the distribution is also a matter of historical geography. China's origins lie in the fertile farmlands of its great river basins, the Huang (Yellow) and Yangzi-Chang in the middle, the Liao-Songhua in the north, and the Xi (West)-Pearl in the south. Here, more than 5,000 years ago, arose the communities that would one day be collocated into a cohesive state, the core of an empire that eventually reached far into Asia in all directions. To this day, and despite all the dramatic events along its Pacific Rim, China numerically remains a dominantly rural society. Behind the façades of Hong Kong and Pudong lie vistas of teeming rural countrysides where farmers by the millions bend to the soil each day as they have for millennia. Their produce, taxes, and tribute built the state; they were persecuted, conscripted, expropriated, dislocated, starved. But in China's great river basins they survived, laying the foundations of a demographic map that has changed little over thousands of years.

Today their world is changing. Pacific Rim China is rapidly urbanizing, and its interior cities are growing fast as well. Wealth is concentrating in this New China, but poverty reigns in the rural provinces, and the gap is widening. Would-be workers from the villages walked and biked by the millions toward the jobs in the shadows of all those new skyscrapers, forcing the authorities to stem the migration and to reverse it. In the headlong rush to modernize, China's planners victimized the farmers once again.

Politically powerless under conditions of endemic corruption, subject to pollution of air and water from new mines and factories in their midst, and disconnected from the forward rush of the New China, the country's farmers —still the majority—have difficult lives and suffer risk and uncertainty. Improving this situation, reducing the income gap and giving villagers a stronger political voice will be China's great domestic challenge as its international stature grows. Failure could endanger the entire national enterprise.

China probably is the oldest continuous civilization on the planet (Egypt might contest the claim), and its sequence of dynastic rule began in the area where the Huang (Yellow) and Wei Rivers meet, not far from the modern capital on the North China Plain. The country alternately fragmented and reunified, absorbed and assimilated outsiders and conquered neighbors, but reached an apogee during the Han Dynasty (206 BC–AD 220), when the Roman Empire thrived at the other end of Eurasia, when the city of Xian was the "Rome of China," and when the Silk Road was a busy trade route. To this day, Chinese call themselves the "People of Han." The Ming Dynasty (1368–1644) was another formative dynasty, but China reached its greatest imperial dimensions during the final traditional dynasty, the Qing (Manchu), which commenced in 1644 and ended in chaos in 1911. The Qing rulers conquered much of Indochina, Myanmar, Tibet (Xizang), Xinjiang, Kazakhstan, Mongolia, eastern Siberia, the Korean Peninsula, and the islands of Sakhalin and Taiwan (Fig. 7-2). Current claims to Taiwan and Tibet and latent claims to

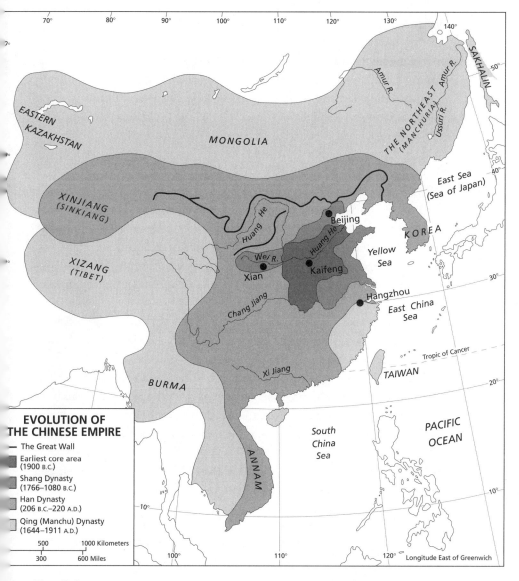

EVOLUTION OF THE CHINESE EMPIRE

— The Great Wall

Earliest core area (1900 B.C.)

Shang Dynasty (1766–1080 B.C.)

Han Dynasty (206 B.C.–220 A.D.)

Qing (Manchu) Dynasty (1644–1911 A.D.)

500   1000 Kilometers
300   600 Miles

Fig. 7-2

Russia's Far East and India's northeast are based on these imperial boundaries, but disaster was in the offing. The nineteenth century was calamitous for a China weakened by overextension, European colonial invasion, and Japanese intervention. British merchants undercut China's economy; opium flooded into China from India and destroyed the very fabric of Chinese society. When the Qing rulers tried to stop this British import, they were defeated in what is called the First Opium War (1839–1842) and the breakdown of Chinese sovereignty was under way. When the Qing Dynasty collapsed in 1911, the empire's territorial losses were already enormous, but worse was to follow as nationalists, communists, and the Japanese fought a three-way war that ravaged the country.

Following the defeat of Japan in the Second World War, China's civil war resumed and Mao Zedong's communists drove Chiang Kai-shek's nationalists from the mainland to Taiwan, where they overpowered the locals and set up their own "Republic of China." In Beijing on October 1, 1949, Mao Zedong proclaimed the birth of the People's Republic of China.

CHANGING THE MAP

China's twentieth-century convulsions were not unlike those of pre-Yuan Dynasty times: the chaotic end of a dynasty amid a breakdown of order, regional fragmentation and civil war, interference from outsiders (in this case not the Mongols or the Manchus but the Europeans and the Japanese), decades of strife ending when a powerful new leader took control. Indeed, after only a few years of Mao's rule scholars began to draw parallels between earlier dynasties and this new communist one. Mao might not have offspring waiting in the wings, but the intrigues and conspiracies marking communist rule and succession were worthy of the dynastic palaces of old.

Mao inherited a weakened China that had lost Taiwan, had British and Portuguese colonists on its doorstep in Hong Kong and Macau, had been forced to accept unfavorable boundary delimitations by more powerful neighbors, and had only nominal influence in far-flung reaches of the state. All these disadvantages would be dealt with eventually, but first the communist regime wanted to address the fate of the millions of landless people and serfs whose lives were indescribably miserable. For all its historic grandeur and cultural achievements, China remained a country with rural areas where floods, famines, and disease could decimate entire populations without any help from the state, where local lords could (and often did) repress the people with impunity, where children were sold and brides were bought. To deal with these horrors, Mao mobilized virtually every able-bodied citizen.

Land was taken from the wealthy, even from the small subsistence farmer. Farms were collectivized, dams and levees were built with the bare hands of thousands. Food distribution was improved, public health enhanced, child labor curtailed. But in the process, China's communist planners also made terrible mistakes. The so-called Great Leap Forward, started in 1958 as the final transformation of China's social life, required the reorganization of the peasantry into communal brigades to combine local industrialization with more productive farming. Peasants were forced into collectivized villages, families were torn apart, and the historic rhythm of agriculture was severely disrupted. Between 20 and 30 million people had died of starvation by the time the Great Leap Forward was abandoned in 1962.

Meanwhile Mao had moved to reassert Han Chinese influence in the country's distant frontiers. Some major losses of the late Qing period could not be reversed, but others could be prevented. Chinese armies entered mountainous Tibet (Xizang) soon after Mao took control in Beijing, and when Tibetans rose against the Chinese in 1959, their revolt was ruthlessly crushed and the Buddhist Tibetans' supreme leader, the Dalai Lama, was ousted. Tibet's historic, fortresslike monasteries, around which Tibetan life was organized, were emptied and much of the fragile society's cultural heritage was destroyed. In desert Xinjiang, peopled by Uyghurs and other Muslim groups with ties to Turkestan, Mao began a program of vigorous Sinicization, the imposition of Han rules and norms. In 1950, Chinese made up only about 5 percent of Xinjiang's population. Fifty years later, the proportion approached 50 percent (of more than 20 million). Here, too, the Chinese faced opposition from its colonized minorities, opposition that more recently has become enmeshed in the global Islamic-terrorism issue.

Two other Maoist initiatives stand out in any assessment of his impact on China's social geography. The first is demographic. Mao often referred to any control of population growth as a capitalist plot to weaken communist societies, and, like his cohorts in Moscow, he encouraged mothers to have numerous children (in the 1960s, Soviet women bearing ten children or more were designated "Heroines of the State"). As a result, following the end of the Great Leap Forward, population growth in China during the 1960s and 1970s was as high as 3 percent, and while the official figures are unreliable, China was adding as many as 20 million per year, rushing toward 1 billion. When Deng Xiaoping's pragmatists took over following Mao's death, one of their priorities was to get this spiral under control, and they instituted their infamous "One Child Only" policy to achieve this. Today, China's official growth rate is 0.7 percent, which, on a base of more than 1.3 billion, still results in an annual increase of 9 million.

The second Maoist initiative was to try to erase the legacies of China's most influential philosopher and teacher, Confucius (Kongfuzi or Kongzi in modern Chinese). Born in 551 BC, Kongfuzi's teaching dominated Chinese life for over 20 centuries. Appalled at the suffering of ordinary people, he encouraged the poor, taught the indigent, questioned the divine ancestries of aristocratic rulers and poured forth a mass of philosophical writing that became guiding principles during China's formative Han Dynasty (206 BC– AD 220). Over time, a mass of Kongfuzi literature evolved, much of which he never wrote but might have approved. At the heart of it lay the Confucian Classics, 13 texts that became the basis for education in China for 2,000 years. From governance to morality and from law to religion, the Classics were Chinese civilization's guide. State civil service examinations, through which poor as well as privileged could gain access to positions of influence, were based on the Classics.

Kongfuzi championed the family as the foundation of Chinese culture, which was incompatible with basics of the Great Leap Forward. He also prescribed a respect for the aged that was a hallmark of Chinese society, but contradicted tenets of the Great Proletarian Cultural Revolution. The communist regime attacked Kongfuzi on all fronts, but Beijing miscalculated. It proved impossible to eradicate two millennia of cultural conditioning in a few decades. After Mao's death, public interest in Kongfuzi surged despite the modernization of education that had been started by the communists, and when you walk into a bookstore in China today you will see shelves sagging under the weight of the Analects and other Confucian writings. Westerners are paying renewed attention too. The last British governor of Hong Kong, Christopher Patten, begins all but one chapter of his reflective book with a quotation from the Analects (Patten, 1998). The spirit of Kongfuzi will pervade physical and mental landscapes in China for generations to come, and understanding this is a key to interacting with the culture.

THE MODERN MAP OF THE CHINESE EMPIRE

Make no mistake: China remains a modern-day empire. When the Soviet Union collapsed, numerous scholars (to my knowledge, none of them geographers) wrote treatises about the fall of the "last empire." More recently one of my colleagues wrote a book entitled *The New Imperialism* suggesting that the United States is the world's last-gasp empire (Harvey, 2004). But the enduring empire is China. Even after the territorial losses sustained during and following the disintegration of the Qing Dynasty, Beijing was left with an imperial realm extending from the Korean border to Turkestan and from

Mongol steppes to Tibetan valleys. First the communists, then Deng Xiao-ping's pragmatists organized this vast contiguous empire with its numerous minority peoples into a spatial framework that would accommodate, at least in principle, their respective objectives.

China's political map has changed repeatedly in recent years, and it will undoubtedly change again. Some of the changes made by the post-Mao administrations reflect a desire in the capital to directly control critical parts of the country. Other changes have been made to facilitate the wedding of communist politics to capitalist economics. Additional modifications have come with the changing status of former European colonies now reabsorbed into the motherland. Should the Taiwan issue be unexpectedly resolved, the map will change again. Meanwhile, new or newly prominent names make their appearance: Shenzhen, Pudong, Xianggang (for Hong Kong). China's geography is a work in progress.

As Figure 7-3 underscores, China's political framework in 2005 had four layers of administration. At the top are four central-government-controlled

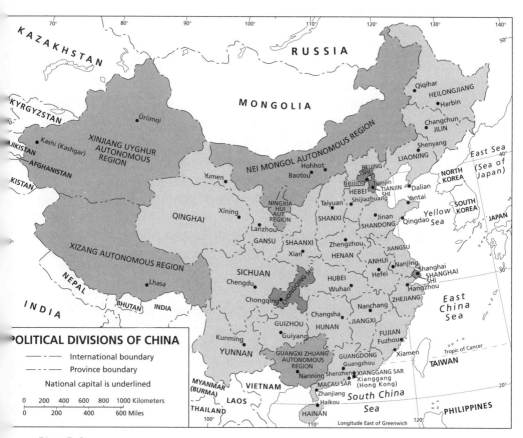

Fig. 7-3

municipalities or *shis*, really the crucibles of Chinese culture and power: Beijing, the capital; Tianjin, its neighboring port; Shanghai, at the mouth of the
Yangzi; and Chongqing, upstream near the great Three Gorges Dam. On the
map, Chongqing looks much the largest, and territorially it is: herein lies a
geographic tale. The city of Chongqing is no giant, but its hinterland lies in
the densely peopled eastern part of the Sichuan Basin, a cluster of more than
100 million people clearly visible on any population map of China, even a
smaller-scale one. When it was decided to make Chongqing a shi, its municipal boundary was delimited far into its hinterland, adding as many as
20 million people to the "municipal" population and putting them under
Beijing's direct control. That made Chongqing, suddenly, the "world's largest
city," as its brochures proudly proclaim. But take this with a grain of salt.
The suburbs do not generate much commuter traffic.

Next come China's 22 provinces. Here is another similarity to the United
States: China's smaller provinces lie in its east and northeast, and its larger
ones, territorially, lie in the west. But here the similarity ends. Look at the
populations of these "small" eastern provinces, and any image of comparative smallness disappears. The four provinces facing the Bo Hai Gulf contain
more people than the entire United States. Thus it is worth noting the locations and populations of some of China's prominent provinces (everyone in
coastal China seems to know about New York, Texas, and California). Even
after "losing" 20 million people to Chongqing city, Sichuan is a province
with nearly 90 million people, roughly the population of Germany, astride
the river that begins its course named the Chang and ends it (near Shanghai)
as the Yangzi. One of China's agricultural breadbaskets and now the site of
the world's largest hydroelectric project, linked to the coast by a major waterway and flanked by one of China's four shis, Sichuan lies at the head of
a wedge that is carrying the economic successes of the Pacific Rim into the
country's interior. This momentous development heralds the transformation
of China's interior, the first step in reducing the regional disparities between
the booming coast and the lagging center.

Two adjacent provinces, landlocked Henan and peninsular Shandong,
each have populations in Sichuan's range, approaching 100 million in 2005
(that's more than California, Texas, New York, and Florida combined!).
Along with neighboring Hebei Province (70 million) they anchor China's
core area, including the capital and the key port of Tianjin. But the provinces
whose names are synonymous with the economic rise of China's Pacific Rim
are coastal Jiangsu and Zhejiang on either side of Shanghai; Fujian directly
opposite Taiwan and for centuries the source of "overseas Chinese" who em-
igrated to Southeast Asia and whose wealth returned to propel the modern

Chinese economy; and Guangdong in the south, where the Pearl River Estuary is evolving into one of the world's greatest urban-industrial complexes incorporating Hong Kong, Shenzhen, Guangdong, Zhuhai, and Macau.

It was Deng Xiaoping's notion to apply his new economic policies in this coastal zone, where market economics and communist politics would coexist without contaminating the rest of the country. Accordingly, the regime introduced a complicated but effective system of so-called Special Economic Zones (SEZs), which in effect were a series of port cities and coastal areas where foreign technologies and investments were welcomed and where investors were offered capitalist-style incentives. Low-wage labor could be hired, taxes were low, leases were simple, and products could be sold on foreign as well as domestic markets. Even Taiwanese enterprises could operate here, though under certain restrictions. The results were spectacular. American and other foreign companies flocked to China's Pacific Rim, jobs by the millions were created, goods poured from newly built factories, and long-dormant coastal cities like Shanghai, Xiamen, Shantou, and even Hainan island's Haikou were transformed by modern office towers, highrise apartments, business parks, and factory complexes. Shenzhen, the SEZ adjacent to Hong Kong, led the way in terms of growth and production, but the whole frontage on the Pearl River Estuary witnessed unprecedented urbanization, industrialization, and modernization. Visitors who have not seen China since the 1980s gasp at this landscape of economic power, its ultramodern bridges, tunnels, highways, trains, airports. It is worth remembering that much of this is happening in China's tropics. Some economists persist in equating tropical environments with "underdevelopment."

This momentous regional growth is not what Mao Zedong anticipated or, if certain Chinese historical geographers are correct, would have preferred. Mao wanted to make China's Northeast (the historic home of the Manchus who had led the Qing Dynasty) the industrial heartland of the People's Republic. The Japanese had occupied this area in the 1930s, and had set up a puppet state called Manchukuo there. China got most of it back, though the Russians held on to a large sector taken from the Qings, and Mao planned to integrate its three provinces tightly into the Chinese command economy. The Northeast had large coal reserves and sizeable oil fields as well as a wide range of minerals including iron, aluminum, manganese, copper, molybdenum, magnesite, lead, and zinc, extensive forests stood ready for exploitation on the slopes of the far north, the Japanese had laid out a substantial infrastructure, the area lay close to the capital, and there was plentiful labor.

The southernmost Northeastern province, Liaoning, was best endowed in every respect and the basin of the Liao River was agriculturally productive as well. Its port, Dalian, was the outlet for the whole Northeastern region. By the end of the 1950s, huge state investments in mines, factories, and transport infrastructure had made such cities as Anshan, Fushun, Shenyang, and Harbin industrial giants, and the provinces of Liaoning, Jilin, and Heilongjiang drew hundreds of thousands of workers to their manufacturing plants.

But virtually nothing produced here, from locomotives to toys and from appliances to furniture, reached, or could compete on, world markets. China's communist planners, like their counterparts in the Soviet Union, assigned production quotas to mines and factories whatever the cost, and even if a locomotive could have been bought at half the price in Canada, the manufacture of railroad equipment continued in Liaoning. China was a closed society, almost a closed system; factories produced as a matter of state policy and workers were secure.

Today, the Northeast is China's tragic rustbelt. The reformers who took over in Beijing after Mao's demise and who opened China's economic potential to the world could not combine market economics with state support for loss-making industries. As government subsidies declined, a few factories in the Northeast managed to adjust and to compete in the new economic order. But most were unable to do so, and plant after plant closed, leaving their workers unemployed and in desperate straits. Beijing tried to mitigate the situation through social programs and some continuing subsidies, and in the early 2000s there were signs that the Pacific Rim effect was reviving the economy in and around Dalian, but thousands of Chinese are now across the border in Russia's Far East as traders, selling cheap goods made in small factories amid the shuttered industries of yesteryear. In the Northeast and in the provinces of the interior, China's transformation has come at a high price.

The third political-administrative level is a relatively new one: the so-called Special Administrative Region (SAR). This category was established to recognize the particular circumstances of Hong Kong's transfer from British dependency to Chinese administration in 1997, and in anticipation of Beijing's takeover of Macau from Portugal in 1999 (a few optimistic observers see it as facilitating the reabsorption of Taiwan at some future time). The notion of "One Country, Two Systems," was much in vogue during the week of Hong Kong's transfer. Hong Kong would become part of China, but it would retain trappings of democracy not available to citizens in China itself. As a member of ABC's television staff I interviewed several British as well as Chinese officials in Hong Kong as the day approached, and it was clear to me that the phrase held different meanings for the two sides. British representatives envisaged a durable democratic system in the ex-colony (which, of course, the British themselves did not nurture until the very end of their tenure, when the so-called Basic Law gave Hong Kong its own constitution). Eventually, they said, the territory's chief executive, at first appointed by Beijing, would be elected directly by Hong Kong's voters. Chinese officials seemed to see it as facilitating the transition but not as a long-term prospect. People would see, they said, that there were more important things in life than democratic politics. The elections of September 12, 2004, appeared to confirm their views. Although the Beijing regime had made it clear that no direct election of the SAR's chief executive was in prospect, and pro-Beijing activists in Hong Kong had tried crudely to tarnish the reputations of prodemocracy candidates, the vote went in favor of pro-Beijing parties and suggested that, indeed, many voters were more concerned about economic conditions than political freedom. The marginalization of democracy appeared well under way, and the slogan of 1997 may turn out to be a mere cliché.

Look at Figure 7-3 one more time, and you will see that vast segments of the Chinese empire are categorized as Autonomous Regions (AR), the fourth level in the country's administrative hierarchy. The Soviet Union had its supposedly autonomous Soviet Socialist Republics: China has its supposedly Autonomous Regions. The terminology may reflect a history of good intentions, but the reality is rather different. Established originally to recognize non-Han minorities living under Chinese control, the five ARs have been exempt from certain national laws (for example, minorities were not subject to the One Child Only regulations instituted under Deng Xiaoping), but they have no autonomy in the practical sense of that word. Xizang (Tibet), formally annexed in 1965, functions as a colony of China, and even though Beijing's harsh rule was relaxed after 1976 and much looted treasure was returned, no expressions favoring independence are tolerated. The Xinjiang-Uyghur AR, where China has still-important oil reserves and where its nuclear weapons, rocket, and space programs are located, is being rapidly Sinicized even as the Uyghurs and other Muslim minorities are tightly controlled. A small pro-independence movement has engaged in occasional acts of sabotage, to which China has responded forcefully; a new railroad to the traditional Muslim city of Kashi (Kashgar) in the far west symbolizes Beijing's determination to confirm its mastery in this Autonomous Region. In the Nei Mongol AR, too, massive immigration has changed the ethnic balance. Originally established to recognize the rights of the several million Mongols who live on the Chinese side of the border with Mongolia, its cultural landscape is now dominantly Chinese. Near the Mongolian border, Mongols still traverse the steppes with their tents and goat herds, but elsewhere this is now a region of farm villages and industrial towns—and Chinese outnumber the Mongols by more than four to one.

As the map shows, there are two other Autonomous regions in China, the Ningxia Hui AR adjacent to the Nei Mongol AR and the Guangxi Zhuang AR in the far south, between the Vietnamese border and Guangdong Province. The Ningxia Hui AR is of interest because it is the historic base of the Hui, who are descendants of Chinese who were converted to Islam when it penetrated western China during the seventh and eighth centuries. The Guangxi Zhuang AR, with nearly 50 million inhabitants, is by far the most populous of the five ARs. The first half of its tongue-twister name refers to the Chinese of the east, and the second half to the Zhuang, a people with ethnic links to the Thai, who are dominant in the interior west. Numerous scattered minorities continue to speak their traditional languages here, including the Yao, Miao, and Dong.

To be sure, this girdle of Autonomous Regions around Han China does not begin to reflect the range or the distribution of minorities throughout the empire. From the Koreans in Jilin to the Li and Miao on Hainan Island, and from the Yi and Dai of Yunnan to the Tibetans of Qinghai, minorities representing about 50 ethnolinguistic groups, most with historic ties across China's borders, live under Beijing's rule. For minorities outside the officially designated ARs, reserves have been set aside in which minority rights are recognized, but often these are also poor and erosion-afflicted domains. On the one hand, the historic strength of Chinese traditions and more recent economic developments have acculturated millions of minority people to Han ways of life, but on the other, millions more resist transculturation and seek to sustain Buddhist, Muslim, or other ways of life.

Certainly the numbers are in favor of the Chinese empire. Few minorities are as clearly defined culturally as the Uyghurs or the Tibetans, and in many parts of southern China, where most of the non-Han ethnolinguistic groups are located, the cultural border between minority and Sinicized is vague and mobile. By some estimates, approximately 8.5 percent of China's population may be categorized as minority—less than half of Russia's more than 19 percent. But keep in mind China's human numbers: 8.5 percent of 1.3 billion people amounts to over 110 million. In actual terms, that is almost four times as many minority people as Russia rules in its vast territory. Like their Soviet counterparts, China's communist rulers inherited an empire from their predecessors, and the Chinese imperial mindset survived the transition.

THE MENACE OF MISCALCULATION

I hope that you stayed with me as we wandered across the political and cultural map of China, because an understanding of that map will someday be a key to avoiding the kinds of miscalculations that induced President George W. Bush to compare Iraq to Germany in the context of bringing democracy to Baghdad. Germany's ethnic, linguistic, and cultural homogeneity and the nature of its neighbors were simply not comparable to traditionally fragmented and multicultural Iraq, and Iraq's regional neighborhood was, to put it mildly, unlike Germany's. Bringing democracy to Germany and to Japan were enormous American achievements that will rank among the proudest in United States history, but the geographic lessons arising therefrom were obviously forgotten in Washington, with horrific consequences. President Bush, however, was not alone in having been misled. In the July 25, 2004, edition of the *New York Times Book Review* section a group of eminent reviewers, discussing the issue of imperialism, proclaimed this to be an essentially Western phenomenon (apparently not only China but also Japan was off their cartographic radar); John Lewis Gaddis referred to the United States as "the only empire left." Obviously none of the discussants was a geographer, but the exchange was on a par with what must have been happening at the White House in the early years of the Bush presidency.

As Ross Terrill points out in his book *The New Chinese Empire* (Terrill, 2003), not only is modern China the product of empire, its expansionist objectives continue. We have already taken note of Taiwan, northeast India, and other actual and latent claims; there is also the question of Mongolia, a part of China during Qing times and now experiencing a strong resurgence of Chinese influence, hitherto in the economic arena but potentially in additional contexts as well. In offshore waters, China is contesting with Japan, Vietnam, the Philippines, Malaysia, and Indonesia the ownership of islands whose acquisition would extend Chinese jurisdiction over vast expanses of the South China Sea. In short, China's territorial drive is far from over.

While many countries have territorial issues with neighbors, these tend to take on greater significance when the claimant is already a giant. More to the point is China's role in competition with the United States for influence and power in the western Pacific from Japan to Australia and from the Philippines to Myanmar. The United States has been the long-term stabilizing force, its postwar relationship with Japan fostering democracy there and creating the setting for one of the twentieth century's great economic successes, its military presence in South Korea protecting one of the Pacific Rim's early economic "tigers" while it prospered and advanced toward democratic governance, and its special relationship with Taiwan precluding a Tibet-like reannexation by Beijing (and nurturing still another economic tiger). America's military presence in the Philippines until 1991, abandoned when Mount Pinatubo's giant eruption destroyed its air and sea bases on Luzon Island even

as the Philippine Senate was weighing continuation of the United States presence, dissuaded China from a greater aggressiveness in its now-renounced
claims to all of the South China Sea. And Washington's close relationship
with Singapore has been another part of this geopolitical framework.

In this new century, however, the picture is changing. Late in 2004, President G. W. Bush announced plans to withdraw United States military forces
from overseas bases including those in Japan. The Japanese, meanwhile,
were bolstering their antimissile capacity in the face of North Korea's nuclear program and rocket tests. United States troops in South Korea were
to be partially relocated from the shadow of the DMZ and partially withdrawn, possibly to Guam. Taiwan, its economy in difficulty, was clearly a
lower priority. Meanwhile the Chinese, always complaining of the asymmetry between the United States presence in East Asia and the Chinese absence
from North America's Pacific Rim, scored a coup when Panama awarded
a contract to a Hong Kong company to operate and modernize the ports
at both ends of the strategic Panama Canal, recently vacated by the Americans. And be prepared for other evidence of China's growing presence in this
hemisphere. China has recently been forging closer ties with Venezuela as
well as Grenada and Dominica, formerly supporters of Taiwan. For China,
the Caribbean is full of opportunity.

When Japan emerged in the third quarter of the twentieth century as
the first economic tiger on the Asian Pacific Rim, buying ever-larger quantities of raw materials from ever-farther sources, it was part of the American success story: stability and democracy had enabled a defeated enemy
to become the world's second-largest economy. Wealthy Japanese entrepre-
neurs bought Western assets ranging from movie studios to golf courses,
art works to historic mansions. Australian schoolchildren by the tens of
thousands took Japanese language courses as Japanese companies bought
Australian commodities by the shipload. But the Japanese economy faltered, and today it is China that is on the rise on the Asian perimeter. From
the mines of Australia to the forests of Myanmar and from the natural gas
of Malaysia to the oil fields of Brunei, China is gobbling up unprecedented
quantities of raw materials. In the process, China's political clout in these
regions grows correspondingly. International observers marvel at the skills
of a modern generation of Chinese business executives and diplomats who
are changing the political as well as the economic climate on the Pacific
Rim—and not just in Southeast Asia but in Japan and South Korea as well.
Today, for all the residual postwar anger between the two countries, Japan
imports more goods from China than from the United States, an almost
inconceivable situation just a decade ago. And China has become South

Korea's largest trade partner after being on nonspeaking terms with Seoul for decades.

Listen to Southeast Asians from Thailand to Indonesia today, and you hear an oft-repeated refrain: China is a potential bulwark against an America whose actions and motives are troubling. The growing Chinese presence combines economic stimulus with political reassurance. Unburdened by human-rights or environmental concerns, China trades actively and increasingly with Myanmar's military junta, ensuring the generals' security (raise this issue, and you will get questions about democracy in Saudi Arabia). Talk about North Korea's nuclear threat, and it is clear that fears of an Iraq-style intervention at the Pacific end of the "axis of evil" play into China's hands. China's star on the Pacific periphery is rising, and geopolitical realities are changing.

None of this, of course, adds up to a cold war—yet. But the perceived twenty-first-century American penchant for unilateralism, impulse, and proselytism is creating conditions that could contribute to its emergence. During the Cold War of the twentieth century, each side was convinced of the other's nefarious designs, and each side saw the other as morally wrong: communist perceptions of the inequities of capitalism were countered by capitalist notions of atheism and repression in the "evil empire." So it comes as an unpleasant surprise to many Americans that the one-time guarantor of global stability is now viewed by many as the more likely protagonist in a Transpacific Cold War.

Are the United States and China on such a collision course, and if so, what are the respective causative and mitigating factors? Let us look again at the political geography of China. On the perilous side, there are China's expansive past and imperial present, its communist-authoritarian governance, its dreadful human rights record, its demographics (the one-child-only policy is creating a surplus of tens of millions of males), its world's-largest military, its fast-rising nationalism, its growing demand for global raw materials including oil, its unsettled relations with Japan, its designs on Taiwan, and its problematic role with regard to its communist neighbor North Korea with its terrorist history and nuclear ambitions. On the mitigating side, China, unlike the Soviet Union, does not overtly seek to export its communist system or ideology except to SARs and Taiwan, maintains a strictly secular society, shares with the United States a concern over Islamic terrorism, has opened its doors to economic development, has settled some territorial issues with neighbors, and has withdrawn far-reaching maritime claims.

China is ascending in a world dominated by a superpower that, during the twentieth century, combined unmatched altruism toward former enemies with still-inexplicable Cold War foreign-policy decisions. Even as the United States brought democracy to (West) Germany and Japan and massive financial assistance to Europe, held the nuclear-armed Soviet Union at bay and opened the door to China's reentry into the world, Washington, like Moscow, supported repressive regimes from Middle America to Monsoon Asia that made Myanmar's generals seem like model democrats. The end of the Cold War led to the end of many of those dictatorships, but the new threat of Islamic terrorism, in part the direct result of an earlier foreign-policy error in Afghanistan, created a new set of circumstances only the outlines of which are visible today. Although there is much concern about the supposedly novel unilateralism in American policy today, this is nothing new. The United States may not have signed on to the Kyoto protocols or other recent treaties, but neither did it ratify the much older UN Convention on the Law of the Sea (UNCLOS) or a number of other international agreements with pre-2000 datelines. What is new, however, is worldwide American interventionism in response to the 2001 terrorist attacks in New York and Washington, a byproduct of which is an inconsistent campaign to install or advocate democracy in countries ruled by authoritarian regimes (inconsistent because it exempts stronger regimes such as China as well as others whose resources are crucial to the American economy, such as Saudi Arabia). In this context the 2003 intervention in Iraq, with its twin aims to depose a ruthless dictator and to establish a democratic government, had far-reaching consequences. To the generals in Myanmar and the despots in North Korea, China's geopolitical ascent in the Asian perimeter was a welcome and timely development. To other countries in the region whose human-rights records or democratic institutions might not meet American criteria, China's deepening economic involvement provides a counterweight to United States pressure.

Is all this enough to precipitate a cold war? Not yet, but if Chinese public opinion is reflected by what is heard in college classrooms, read in the press, and seen on the Internet, the prospect has risen in the nation's consciousness. American criticisms of China's human-rights deficiencies are met with angry denunciations of the failures of American race relations; American commentaries on China's undemocratic system of government are countered with critical analyses of the shortcomings of democracy in the United States. Public support for United States intervention in Afghanistan turned into widespread opposition to the invasion of Iraq, and in my experience disapproval of the involvement of South Koreans in that campaign is nearly universal in China. In the process, politics has begun to displace economics as the chief topic of discussion and debate, which is not a good sign for the future.

The most populous nation on the planet, heirs to a great empire and guardians of one of the oldest continuous cultures, is asserting its place in a world dominated by American superpower. China is the first non-Western power to mount such a challenge, creating a new set of geopolitical circumstances. During the twentieth century, when the United States and the Soviet Union were locked in a Cold War that repeatedly risked nuclear conflict, Armageddon never happened because this was a struggle between superpowers whose leaderships, ideologically opposed as they were, understood each other relatively well. While the politicians and military strategists were plotting, the cultural doors never closed: American audiences listened to the music of Prokofiev and Shostakovich, watched Russian ballet, and read Tolstoy and Pasternak even as the Soviets cheered Van Cliburn, read Hemingway, and lionized American political dissidents. In short, this was an intracultural Cold War, which reduced the threat of calamity. A cold war between China and the United States would involve far less common ground, the first intercultural cold war in which the risk of fatal misunderstanding is incalculably greater than it was during the last.

How can such a cold war be averted? Trade, scientific, cultural, and educational links and exchanges are obvious remedies: the stronger our interconnections, the less likely is deepening conflict. We Americans should learn as much about China as we can, to appreciate its experiences and to understand the historic and cultural geographies underlying China's views of America and the West. China today is a major player on the world stage, internationally involved and globally engaged. Constraining the forces leading to a world-power competition between the United States and China will be the key geostrategic challenge of the twenty-first century.

8

TERRORISM'S WIDENING CIRCLE

On the morning of May 14, 1940, the skies over Rotterdam, the Netherlands' second city and leading port, suddenly filled with warplanes coming from the east. Without warning they rained incendiary bombs on the historic center, the harbor, and other facilities. Hundreds were killed in the first wave of the assault, and when survivors tried to douse the flames with fire hoses they were shot by Nazi paratroops who had landed at the airfield. I watched the conflagration with my father from a skylight in my family's suburban house, his arms around me at the top of a tall ladder. Nightfall colored the sky over the city a vivid, menacing red as explosions and collapsing buildings sent columns of black smoke towering over the ruined city.

The attack was a signal to the Netherlands' government, still resisting German aggression, to capitulate or risk further death and destruction. To my recollection, no one later referred to it as a terrorist act, although the great majority of the casualties were civilians. It was an act of war between states, and there was no doubt regarding the identity or the location of those responsible. After the end of the war, my father told me that some of his acquaintances in Rotterdam, soon after the bombing, had actually blamed the Dutch government for it. If only the government had recognized the inevitable and allowed the Nazis to overrun Holland without opposition, Rotterdam wouldn't have happened.

Over 60 years later, as my parents (then aged 96 and 94) and I watched in horror as the September 11, 2001, events unfolded in New York and Washington, this boyhood experience came rushing back. As the second plane struck the towers we turned simultaneously to each other and said "Rotterdam." But we were mistaken. The identity and location of the perpetrators was by no means immediately clear. It was not a formal act of war, but of terror. Nevertheless, in United States newspapers there subsequently appeared letters from readers asking "what it is we (Americans) did to cause such retribution." That was Rotterdam all over again.

WHAT IS TERRORISM?

If Rotterdam was an act of war and not terrorism, was Dresden any different? That German city was targeted by Allied bombers beginning in mid-February 1945 for reasons similar to Rotterdam in 1940: to persuade a government to capitulate and thus shorten the war. Perhaps as many as 100,000 civilians were killed, but the sustained campaign had no effect on Nazi policy. How about Hiroshima and Nagasaki? The argument over casualties continues: would more American and Japanese soldiers (as well as civilians) have died in the inevitable invasion than perished in the two cities destroyed by United States atomic bombs? Is there such a thing as state terrorism, and if so, do these four cities exemplify it? Certainly the Palestinians are in no doubt about the reality of state terrorism. The Palestinian press regularly accuses Israel of it when Israeli warplanes and tanks in pursuit of Arab terrorists kill civilians in Gaza and the West Bank.

The definition of terrorism is no minor issue. It raises heated arguments in the scholarly literature, and it has important legal and fiscal (notably insurance) ramifications. In a book entitled *The Geographical Dimensions of Terrorism*, terrorism is defined as "intimidation through violence," but without further qualification (Cutter et al., 2003). Neither is the definition written by the United Nations Panel on Threats, Challenges and Change satisfactory: Terrorism is "any action intended to kill or seriously harm civilians or non-combatants, with the purpose of intimidating a population or compelling action by a government or international organization" (Annan, 2004). A more considered view defines terrorism as "an unprovoked attack against civilian noncombatants away from any theatre of war by men or women not working for a power openly at war with the victims' country" (Harvey, 2003). The problem with such carefully qualified definitions is that further definitional problems arise: what is "unprovoked?" Terrorists claim a host of provocations, from cohorts' deaths to historic injustices. Does "openly at war" mean that a rebel movement's declaration of war on a government— which has occurred repeatedly—is enough to negate that clause and thus the charge of terrorism? To such questions, there is no simple answer.

A RISING TIDE OF GLOBAL TERROR

"One person's terrorist is another person's freedom fighter" goes a common refrain when it comes to violence to achieve political goals. When the Irish Republican Army (IRA) claimed responsibility for a bomb attack on Prime Minister Margaret Thatcher at the Grand Hotel in Brighton in October 1984, killing 5 and wounding 31, British public opinion held this to be an act of terrorism whereas many Catholics in Northern Ireland viewed it as justified resistance. When a Palestinian suicide bomber destroys a bus full of civilians on an Israeli street, Jews see an act of terror, many Palestinians a heroic strike at a hated enemy.

In these and many other examples known to all of us, the violence is an outgrowth of national policies. The IRA fought against British rule in Northern Ireland (the issue is closer to being settled today, but no final resolution has been achieved and the potential for further upheaval exists). Palestinians fight against an Israel they resent as a trespasser, invader, and occupier. Corsicans in France, Basques in Spain, Tamils in Sri Lanka, Chechnyans in Russia, Uyghurs in China—in every instance, their battle is against the state that dominates them, and in every case they have carried their violent campaigns to the major cities of their adversaries. Corsicans bombed the city hall of Bordeaux; Basques assassinated politicians and civilians from Madrid to Barcelona; Tamils killed hundreds in Colombo, Chechnyans turned Moscow into a war zone, Uyghurs detonated bombs in Urümqi. None of the long-standing issues provoking these actions is yet resolved, but their protagonists are well known and their geographies are tangible. What the activists want is a redrawing of the map: an autonomous Corsica, an independent Basque territory, a partitioned Sri Lanka, a sovereign Chechnya, a free East Turkestan.

No such geographic clarity marks the form of terrorism that has become dominant in the twenty-first-century world: Islam-inspired violence. A generation ago, acts of terror punctuated political strife in numerous places around the world, affecting not only the locales where it was endemic but also such countries as Canada, Italy, India, and Japan. In the United States, occasional terrorist acts tended to be perpetrated by far-right or neo-Nazi activists, often operating individually. Elsewhere, shadowy groups with diverse grievances and objectives used bombs, guns, hijackings, kidnappings, and other means in pursuit of their objectives, which often seemed to be, simply, to create mayhem. Until the early 1980s, although some cooperation did occur, there was no global network coordinating these groups. But circumstances were evolving. Terrorism was becoming more than a local crime problem to be dealt with by local law enforcement. A new and larger wave of terrorist activity was rising, overshadowing all the local movements and creating a global threat, targeting Western interests worldwide. This terrorist network had religious rather than ideological foundations and exhorted Muslims everywhere to participate. It found a ready market among Salafists (Salafism demands strict emulation of Islam's seventh-century founders) and others who espoused Islamic revivalism in its most fundamental forms.

Even so, little was known about this fundamentalist movement just two

decades ago. Islamic terrorism had for some time been equated with Palestinian Liberation Organization (PLO) activity against Israel and Israeli
interests, ranging from the deadly attack against Israeli athletes during the
1972 Olympics in Munich to the 1985 hijacking of an American airliner
leaving Athens by a group of Shi'ite Muslim terrorists demanding the re-
lease of fellow Shi'ites in Israeli prisons. The widening circle of this violence
sometimes struck United States interests, but its source remained obscure.
In 1982, the name "Islamic Jihad" (Islamic Holy War) began to appear on
intelligence reports, a shadowy organization that claimed responsibility for
four terrorist attacks during 1983 and 1984 on American military and diplomatic installations that killed nearly 320 people. Yet, as one perceptive observer wrote at the time, "there [are] doubts that Islamic Jihad, as a proper
organization, actually exists. No names or locations have been publicly
identified . . . it is widely believed that 'Islamic Jihad' is a convenient cover
name for a host of fundamentalist movements and cells throughout the
Middle East" (Wright, 1986). Speculation as to Islamic Jihad's headquarters
centered on Shi'ite Iran, in part because of Lebanon-based Hizbullah and
its known ties to Tehran. But in truth there was little concrete information about an enemy proving itself capable of inflicting heavy casualties on
American and other Western interests.

Even as such attacks continued, the seeds for a far more serious challenge to
regional security were being sown elsewhere. After the Soviet Union invaded
overwhelmingly (Sunni) Muslim, landlocked Afghanistan in 1979 with the
aim to establish a secular regime in Kabul, the United States made the fateful foreign-policy decision to support anti-Soviet *mujahideen* (strugglers) by
providing them with large quantities of modern weapons as well as money. It
was a Cold War strategy that might be charitably described as shortsighted,
and certainly appeared contradictory: a United States supposedly committed to principles of secular government helped oust a regime attempting to
establish just that, albeit undemocratically. Soviet restraint in the United
States struggle for Vietnam was not rewarded with American moderation in
Afghanistan, even when it must have been obvious to some policy makers in
Washington that a forced withdrawal of Soviet troops would leave Afghanistan a destabilized, faction-ridden, failed state flooded with armaments at a
time when Islamic terrorism was on the rise. Soon the factions that had been
somewhat united during the anti-Soviet campaign were in conflict, with
warlords controlling fiefdoms far from the devastated capital and more than
3 million refugees encamped on the Pakistan side of the border.

The United States, of course, was not the only factor in the struggle for Afghanistan. Another was Saudi Arabia, source of megafunds in support of the

mujahideen and wellspring of Islamic extremism that would find fertile soil amid the cultural chaos that followed the Soviet defeat. For decades, Saudi money had supported numerous religious schools or *madrassas* in Pakistan, schools in which education is confined to intensive study and memorization of the Quran (Koran). These schools had their origins in British colonial times, when Muslims found themselves not only colonized and deprived of their power but also a minority in the dominantly Hindu South Asian realm. In response, as early as the 1860s, Sunni Muslim communities set up such schools. The cleric-teachers of these madrassas were able to issue *fatwas* (legal, interpretive proclamations based on the holy texts of Islam, the Quran or the Hadith, that stipulate how Muslims should live their lives, and thus to protect Muslim communities from undesirable influences). When Pakistan was severed from India and became an Islamic state, the Pakistani madrassa network became a potent factor in national life. Supported by taxes and drawing youngsters by the hundreds of thousands from the Sunni majority (the Shi'ite minority generally disapproved and objected to being taxed for this system), the madrassas provided lodging, food, and religious education mainly for the sons of poor families. During the war in Afghanistan, when several million Afghans took refuge in Pakistan, many of their sons also enrolled in these schools. Years of indoctrination and rote memorization left them poorly prepared for the real world, and the religious perspective with which they graduated was—and is—retrograde.

In the eyes of their Saudi supporters, however, there was—and is—nothing retrograde about Pakistan's madrassas. Indeed, their teachings tend to conform to those of fundamentalist and often extremist Wahhabi dogma that pervades conservative Saudi religious proselytism. The Arabian-born theologian Muhammad ibn Abd al-Wahhab (1703–1792), the founder of this movement, was determined to bring Islamic society on the peninsula back to its most traditional and puritanical form. So strict were his teachings in his hometown near Medina that he was expelled by the locals in 1744, when he moved to what was then the capital of the central territory of Najd, whose ruler was named Ibn Saud (that old capital lies close to the modern one, Riyadh). That was a fateful move, because Ibn Saud liked al-Wahhab's teachings and formed an alliance with him that was facilitated by the fortune al-Wahhab had inherited upon his wife's death. Together they set out on a path of conquest, political and religious, that created a ruling Saudi dynasty over most of the Arabian Peninsula and made Wahhabism the dominant dogma. In 1932, when the Kingdom of Saudi Arabia was officially established, the ruling king made Wahhabism the state religion.

The era of oil production, foreign corporations, modernization, and immigration by millions of Arabs and non-Arabs from Palestine to Pakistan changed Saudi Arabia substantially, but Wahhabism survived. The Saudi royal family had grown by the turn of the century to nearly 5,000 members, and while some had Western educations and others had foreign experience in diplomatic and business fields, still others followed Ibn Saud's traditions and remained staunchly conservative, supporting fundamentalist religious institutions within Saudi Arabia and mosques and schools abroad. In Pakistan's madrassas these conservatives saw the values they embraced, and in conservative clerics they found allies in support of their revivalist cause (many Muslims prefer the term "revivalist" to "fundamentalist," arguing that the former is appropriately forward looking). Thus money flowed to mosques and schools abroad, fueling Islamic revivalism while creating channels to funnel such money to fanatics willing to take up arms, and give their lives, on behalf of extremist causes.

From Pakistan came the products of the madrassas, from Saudi Arabia came money. But there was a third element in the transformation of Afghanistan from isolated backwater to terrorist base. In several ways, the life of Usama bin Laden mirrors that of al-Wahhab: he gained significant wealth through inheritance, in his case from his father; following college graduation he left Saudi Arabia to pursue his causes elsewhere; he returned to denounce his leaders' lack of commitment to Islamic principles; and he was exiled—not just from his local community, but from his country. Bin Laden's final chapter has not yet been written, but if he could write it the story would involve a triumphant return to Saudi Arabia to destroy the monarchy, oust the Americans, and establish a theocracy. During the 1980s, however, Usama bin Laden and the United States had a common goal: to oust the Soviets from Afghanistan. He used his own money and an even greater hoard from Saudi royal and religious donors to propel this campaign, watching his fame as well as his war chest grow in the process. In 1988, even before the last Soviet troops had left Afghanistan, he and a group of associates founded an organization called al-Qaeda. Following the final withdrawal of Soviet forces, he returned to Saudi Arabia, leaving behind a large network of allies in the region. Once back in his home country, he publicly denounced his government for allowing United States troops on Saudi soil during the 1991 Gulf War, when an alliance of countries including Arab states, led by American forces, ousted the Iraqi army from Kuwait. By this time bin Laden had become a prominent figure, and the Saudi regime wasted no time in expelling him from the country, stripping him of his citizenship. Allegedly with the help of a conservative member of the royal family, he found refuge in Arab-ruled Sudan, where he established a number of legitimate businesses to facilitate his now-global financial transactions, but also set up several terrorist training camps. Next he convened conferences attended by terrorist leaders and activists representing an ever-growing number of allied organizations ranging from Jemaah Islamiya (Indonesia) and Abu Sayyaf (Philippines) to Islamic Jihad (Egypt) and Hizbullah (Lebanon). When this became known, the Khartoum regime came under strong international pressure to oust him, and in 1996 bin Laden and his circle of confidants transferred their headquarters back to Afghanistan.

By now bin Laden's role as leader of a terrorist network far more extensive than Islamic Jihad was well understood, and his organization, al-Qaeda, had claimed responsibility for numerous major terrorist strikes, including the first attack (1993) on the World Trade Center in New York City killing six and injuring about one thousand people. Indeed, bin Laden issued a series of fatwas against the West and the United States, in one of them actually declaring "war" on America. But bin Laden also continued in his role as a businessman with global reach. For example, his organization controls what is referred to by counterterrorist operatives as the "al-Qaeda Navy," consisting, press reports suggest, of about two dozen ships. These ships ply the oceans doing legitimate business carrying such freight as cement and bags of beans. "In 1998 one of them delivered the explosives to [East] Africa that were used to bomb the United States embassies in Kenya and Tanzania. But immediately before and afterward it was an ordinary merchant ship, going about ordinary business. As a result, that ship has never been found" (Langewiesche, 2004).

AFGHANISTAN IN THE CROSSHAIRS

When bin Laden returned to Afghanistan, the long-term investments made by Pakistan and by Saudi interests in the madrassas were bearing fruit. An Afghan jihad was in progress, driven by *taliban* ("students") from these schools in Pakistan and led by *ulemas* (teachers) who had turned them into religious fanatics. Also on the scene were an assortment of terrorists of note, including a physician named Ayman al-Zawahiri, who had been involved in the murder of Egyptian president Anwar Sadat and who was to become second in command of al-Qaeda. When the Taliban (the capitalization reflects their role in the holy war) entered Afghanistan from their supportive Pakistani base, Afghanistan was fractured into several dozen fiefdoms ruled by mujahideen warlords whose forces were well armed and who controlled trade and transportation, levying tolls and tribute and taking the country back to feudal times. The skeletal government in Kabul had little power in the city's immediate hinterland and none in the remote provinces. But it

was clear that the areas where the largest ethnic group—the Pushtuns—
were based had some semblance, at least, of stability. Thus when the mainly
Pushtun Taliban seized control of the southern city of Kandahar and moved
toward Kabul, they had considerable support in a population exhausted by
decades of conflict. But even as their campaign to overpower the last resisting warlords continued, the Taliban began to impose upon their subjects the
rules they had learned in their schools: the rules of sharia law. Women were
compelled to wear burqas and prohibited from employment. Many of those
working women "who had lost their husbands, fathers, and brothers in the
war were forced to beg in the streets surrounded by their starving children"
(Kepel, 2002). Entertainment in the form of television, radio, or live music was forbidden; on Friday the soccer stadium in Kabul was used not for
games but for the public whipping of accused drinkers, the amputation of
the hands of thieves, and the execution of condemned murderers. From his
residence in Kandahar, the Taliban's Commander of the Faithful, Mullah
Omar, endorsed such practices and directed Taliban strategy against the remaining resistance. Surrounding him was the growing circle of Islamic terrorists who had converged on the safe house Afghanistan had become. Chief
among them was Usama bin Laden, who would soon begin plotting the 2001
attacks on New York and Washington.

The physiography of Afghanistan and the country's relative location made
it a receptive locale for such purposes. Nearly the size of Texas, compact in
shape except the protrusion of land called the Vakhan Corridor extending
eastward to touch China, mountainous in the center with countless valleys and steep-walled, strategic passes among which the Khyber Pass is the
most crucial, and wedged between Shi'ite Iran to the west and Sunni Pakistan to the east, Afghanistan is the very definition of remoteness, isolation,
and fragmentation (Fig. 8-1). In the east, rugged, forested terrain marks the
border area with Pakistan, and in this cave-riddled topography Usama bin
Laden was able to escape his pursuers, making his way (in all probability)
across the border from Tora Bora and hiding on the rugged, wild, culturally
closed Pakistani side. Here, in what maps call the Tribal Areas (Waziristan),
bin Laden found allies to hide him and fighters who for as long as Pakistan
has existed have defied the government and habitually killed unwanted in-
truders. Even before the American intervention began, it was clear that this
would be bin Laden's (and Mullah Omar's) best refuge, and the plan, if there
was one, to interdict them was ineffective. In any case, Afghanistan's internal
as well as cross-border fragmentation extends to its cultural geography: no
ethnic group in the population of over 30 million constitutes even half the
total. The Sunni Pushtuns may constitute 40 percent of the total (data are

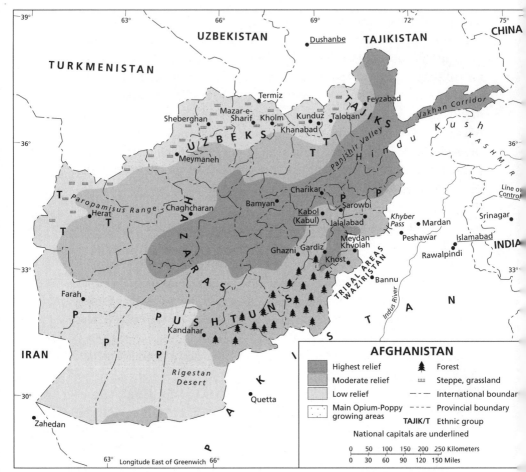

Fig. 8-1

unreliable) and occupy the most densely peopled center and east, where the
capital is located. Across the ranges of the Hindu Kush in the far north are
the Tajiks and the Uzbeks, whose ancestral relatives live in the neighboring
countries named after them. Where the Hindu Kush yields to the plainlands
of the west live the Shi'ite Hazaras, and in the desert south the concept of
Afghanistan as a state is only marginally relevant to the Baluchi.

At the turn of the century Afghanistan's vital statistics ranked it lowest in
the world by several criteria. In terms of child and infant mortality, life expectancy (this is one of the few countries in the world where men live longer,
on average, than women, 46 to 44 years), income, education, medical facilities and services, nutrition, and other indicators Afghanistan was in terrible
shape. Urbanization was barely over 20 percent; roads between cities and
towns were little more than dirt tracks subject to closures due to flash floods,
landslides, and rockfalls. Even the main "highway" between the national

capital Kabul and the "southern capital" Kandahar was a potholed stretch of gravel along which robbers lay in wait. If we had a map of traffic flows in Afghanistan, we would find that more movement took place between Afghanistan's periphery and neighboring countries, for example from Herat to Iran and from Mazar-e-Sharif to Uzbekistan, than within Afghanistan itself (the heavily used Khyber Pass and other eastern passes to Pakistan would confirm this). Most of the four-fifths of the population living in the rural areas subsisted on unreliable food crops or opium-poppy cultivation. A few valleys, such as the Panjshir in the northeast, were well watered and had soils that could be depended on to yield good annual harvests, but the majority of farmers lived a difficult life of uncertain subsistence. To them, the Taliban were a welcome stabilizing force after decades of wartime dislocation. What the Taliban did to their urban counterparts was no concern of theirs.

As Afghanistan descended into medieval misery and the Taliban engaged in despicable acts of oppression and retribution while ravaging the country's cultural landscape (the dynamiting of two ancient, huge, stone-carved statues of the Buddha at Bamyan in the Shi'ite Hazara area raised a worldwide storm of protest and a few expressions of "unease" in the Muslim world), neighboring Pakistan continued to recognize its regime as legitimate. From the United States government and from the Saudi regime came demands for the ouster of bin Laden; from other sources in Saudi Arabia the flow of money supporting the Wahhabists and their terrorist guests continued. And the Taliban had their own source of income: their avowed disapproval of narcotics notwithstanding, Taliban-ruled Afghanistan continued to produce more than 70 percent of the world's heroin. Meanwhile, the global Islamic terrorist network now centered in and funded from Afghanistan had repeated successes, including the killing of numerous Western tourists at Luxor, Egypt in 1997, the bombing of the United States embassies in Nairobi, Kenya and Mombasa, Kenya in 1998, the destruction of apartment buildings in Moscow and other Russian cities in 1999, the assault on the USS *Cole* in the port of Aden, Yemen in 2000, and the attacks on New York and Washington in 2001. The 9/11 attacks led to the United States invasion of Afghanistan and the overthrow of the Taliban regime, but, as noted, both Mullah Omar and Usama bin Laden escaped and the al-Qaeda leader continued to issue taped statements to the public as well as instructions to operatives.

THE WIDENING CIRCLE

In the wake of the American action in Afghanistan and the Allied reconstruction of the country leading to its first democratic election in late 2004, the al-Qaeda network was clearly disrupted, and its strategies, funding, and operations were dislocated. Bin Laden apparently had notions of a global campaign extending from Singapore (where a team of al-Qaeda operatives were arrested) to Bosnia and from Los Angeles (targeted by terrorists detained crossing the border from Canada) to Russia, prominently including his homeland of Saudi Arabia. It is probable that crucial elements of those plans were foiled, but individual nodes in the al-Qaeda network continued their activity and targets in Saudi Arabia came under repeated attack. In 2002, an assault on a Bali (Indonesia) nightclub killed nearly 200 people. In 2003, an explosion at the Davao International Airport in the Philippines killed 21 persons and suicide bombers killed more than 40 people, many of them foreigners, in simultaneous attacks in Casablanca, Morocco. In 2004, bombs hidden in backpacks exploded aboard morning commuter trains in Madrid, Spain, killing more than 190 passengers and affecting the national elections scheduled days later; in the Russian republic of South Ossetia, terrorists killed more than 300 schoolchildren after invading a school in the town of Beslan. In each of these incidents and many others not cited, the affiliations of the perpetrators remained in doubt, and if they were arrested, their nationalities might not reveal the sources of their support or the origins of their instructions. Al-Qaeda remains a potent name in the annals of terrorism, but other terrorist organizations are sharing its mantle. Jemaah Islamiyah, Ansar al-Islam, Al-Jihad, Salafist Group, Abu Nidal Organization, Hizbullah, and dozens of other terrorist outfits infused with Islamic extremism remain linked in a network that will probably strengthen and weaken cyclically depending on successes and failures in the war on terrorism.

What these organizations have in common is their lack of definition. Even Hizbullah, historically associated with the small state of Lebanon, is not an arm of the Lebanese state, nor are its activities confined to Lebanese territory. Hizbullah has long had fiscal and ideological links with Iran, and its terrorist operations have a wide international reach touching many countries (Hizbullah also runs social programs providing food, medical care, and education). Even if the affected countries wanted to, they would be unlikely to have the capacity to constrain the movement. And if they did, the organization would melt away and regenerate elsewhere. In this respect, Islamic terrorism and its organizational networks create entirely new challenges for the international community. Prevention and interdiction, not force and retaliation, are the keys to security.

Islamic terrorism is today's growing challenge, but other forms of terrorism are not extinct—they merely receive less attention in this era of random risk. The United States Department of State's Office of the Coordinator for

Counterterrorism publishes an annual report entitled *Patterns of Global Terrorism* that lists all officially designated foreign terrorist organizations (U.S. Department of State, 2004). The current issue identifies 37 such terrorist organizations of which 23 are Islamic and the other 14 represent such groups as Basque Fatherland and Liberty (ETA) in Spain, the Liberation Tigers of Tamil Eelam (LTTE) in Sri Lanka, the National Liberation Army (ELN) in Colombia, Aum Supreme Truth (AUM) in Japan, the Sendero Luminoso (SL) in Peru, and the Real Irish Republican Army (RIRA) in Northern Ireland. The difference between these uncoordinated organizations and the Islamic network is that none of the former presents an international threat; their aims are local. Even the Palestinian organizations that carry out and support terrorist acts against Israel eventually ceased their hijackings and attacks elsewhere, because it became clear that such practices turned world opinion against Palestinian causes (in this context, the October 2004 bombing of the Hilton Hotel in Taba, in Egypt just across the border from Israel, may represent a reversal of policy or a non-Palestinian action). While al-Qaeda and its allies may not be as tightly and effectively connected as they were before 9/11, their targets and goals are global, not local. Disrupting the coordination among the nodes and cells in the network is the key to security in the years ahead.

GEOGRAPHY OF RAGE

On the walls of many a geography teacher in the Muslim realm hangs a map not unlike Figure 8-2, depicting all areas of the world that are, or were at one time, under Islamic sway. That map shows a contiguous Muslim *umma*, or world, extending from West Africa to Central Asia and from Eastern Europe to Bangladesh, with an outlier in Southeast Asia. It includes most of Spain and Portugal as well as Hungary, Romania, Bulgaria, and Greece, much of India and part of western China. It prompts memories not only of Islam's global reach but also of its early splendor as a culture whose achievements in science and mathematics, architecture and the arts far exceeded those of Europe. Refer to this map, and even "moderate" Muslims tend to become quite agitated. "I remind myself daily of the humiliations we have suffered and the lands and peoples of God we have lost," a teacher at an otherwise modern, well-equipped school in Alexandria, Egypt told me in 1980. "This map is our inspiration, our minimum demand on the world." I was unaware at the time of the portent of those sentiments, versions of which I had already heard in South Asia and East Africa and would hear again in Dubai and Morocco. I thought of this map when Usama bin Laden, in one of his post-9/11 taped

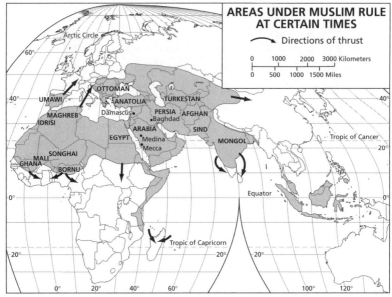

Fig. 8-2

diatribes, referred to the "loss" of al-Andalus, now part of Spain, among his justifications for the carnage he had caused.

This sense of shame and mortification is widespread, if not universal, in the contiguous Muslim world: Muslims once ruled an Ottoman Empire that reached from present-day Turkey to the gates of Vienna and a Mogul Empire that extended from modern Pakistan to Bangladesh. They lost Iberia and they were ousted from almost all of Eastern Europe. They were battered by the Crusaders and colonized by the Europeans and the Russians, whose geometric boundaries of administrative convenience created eventual states of impractical configuration (Iraq was one of those states). They had no voice when, in the aftermath of a war among Western powers, the United Nations carved out a state of Israel in their Middle Eastern midst. And then the West's voracious appetite for oil brought foreign economic, cultural, and political penetration of what remained of the Islamic realm, even of the holiest of lands, the Arabian Peninsula, site of Mecca and Medina.

Thus there is no reason to ask, in the aftermath of 9/11, what it is Westerners in general or Americans in particular have recently done to persuade 15 middle-class Saudis and four other Muslims to send more than 3,000 people to a fiery death in coordinated suicide attacks so terrifying that they unhinged the civilized world. In the annals of history, what was done to Islam and Muslim peoples is proportionately no more dreadful than what was done by European (and Arab) enslavers to Africans, American settlers to Native Americans, Belgians to Congolese, Germans to Jews, and too many other depredations to chronicle. But Africans are not engaged in suicide missions in Brazil, Native Americans are not bombing United States cities, Congolese are not targeting Brussels, and Israelis are not blowing up German commuter trains. If settling historic scores were to have become routine, the planet would no longer be a livable place.

It is often argued that the answer to the question just posed lies in the Israeli-Palestinian struggle and the consequent radicalization of Islamists everywhere, creating a growing cadre of terrorists ready to give their lives in the cause of Allah. But there is little evidence for this. The cycle of violence that has ensnared Israel and the Palestinians has more in common with Northern Ireland, the Basque problem, and Tamil Eelam than with al-Qaeda's campaign, which means that even a two-state solution to the satisfaction of both parties will have little or no impact on the larger issue. No Palestinians are known to have been involved in the planning, funding, or execution of the 9/11 attacks. A majority of Palestinians, their conflict with Israel notwithstanding, would favor a peaceful and territorially fair resolution and would accept Israel as their neighbor. This is not the position of al-Qaeda or its allies, and it is not what that map of the immutable Muslim world reflects.

No amount of behavior modification on the part of the Western world or of the United States can undo what history has wrought, and of course there is no way to return the planet to the circumstances represented by Figure 8-2. Not even a complete cessation of oil exports and the total withdrawal of all Westerners and Western interests from Saudi Arabia would be enough to satisfy bin Laden and his Wahhabist associates: they equate the "moderate" wing of the royal family with the former shah of Iran, and nothing short of a theocracy of the Khomeini variety will do. Indeed, Khomeini himself made a move that reveals the intent of those who espouse the true faith: in 1989 he issued a fatwa that reached beyond the world of Islam, the umma, by proclaiming a death sentence against a British author living in the United Kingdom for a work allegedly containing blasphemy. This pronouncement compelled Muslims to attempt to find and kill the offender, who had to go into hiding in his own country. Not just the Islamic world, but the entire world must countenance the laws of Islam.

Let us look again at Figure 8-2 and compare it to maps of world natural environments, especially climate (Fig. 4-4). It is instructive to note that the contiguous region where Islam prevails today (thus the area shown minus Iberia and most of Eastern Europe; Islam no longer rules but still has strength in India) coincides remarkably with the world's harshest desert climates. Indeed, the harshest forms of Islam seem to prosper in the toughest environments: Saudi Arabia, Pakistan, Sudan; the milder forms of it appear to predominate under more moderate environments in Indonesia, Malaysia, and Bangladesh. This is not to suggest a causative relationship, but it relates to another observation of the spatial character of the faith, which is that its core area remains most fundamentalist while the periphery (not only Indonesia but also Turkey, Morocco, and Senegal) tends to be more temperate. This may be taken to a larger scale: on the Arabian Peninsula itself, coastal Dubai, where women (unlike Saudi Arabia) may drive, and Oman, where some schools admit girls as well as boys, conditions in Islamic society vary. In the Middle East, Saddam's Iraq was a secular state under the Sunnis' harsh rule; its leaders wore business suits or military uniforms, not religious garb. In Lebanon, about one-third of the Arab population remains Christian (down from 50 percent a half century ago), and even some Palestinians remain Christians today. In the Maghreb, the Atlas Mountains form a cultural divide as well as a physical one: the coastal zone with its more cosmopolitan cities and towns and their busy bazaars are a far cry from the Berber interior and its villages and caravans.

Does this mean that moderation has a chance, that urbanization, migration, globalization, and economic interaction will eventually temper the rage that accompanies the revivalism driving Saudi Arabia's angry conservative clerics, their fiscal sponsors at home and their Wahhabist allies abroad? It does not appear so. In the aftermath of 9/11, well-intentioned American colleges began teaching courses about Islam and the Quran, but Islam's holy book, like the Bible, makes discouraging and well as difficult reading. True, it contains hopeful sentences such as "There shall be no compulsion in religion" (Quran, 2:256), but anyone hoping that the contradictions familiar to Bible readers may be fewer in this holy book will be disappointed. "On almost every page, the Koran instructs observant Muslims to despise nonbelievers. On almost every page, it prepares the ground for religious conflict" (Harris, 2004). That bleak assessment actually understates the case. The Quran's angry denunciations of all who qualify as "infidels," its warnings against interaction with "unbelievers" and its promises that "those that deny Our revelations will burn in fire . . . no sooner will their skins be consumed than We shall give them other skins, so that they may truly taste the scourge" (Quran, 4:55) are read by the faithful and elaborated by mullahs in sermons rarely marked by restraint. In Islam, Harris argues, "the basic thrust of the doctrine is undeniable: convert, subjugate, or kill unbelievers; kill apostates; and conquer the world."

Thus the territorial imperative reflected by Figure 8-2 is in fact exceeded by the Quran's stipulations, and in addition it is matched by the faith's position toward those who might wish to change religions from Islam to, say, Buddhism, Hinduism, or Christianity. Numerous Christians have converted to Buddhism; others have converted to Islam, some now-prominent Americans (found fighting for Islamic causes) among them. But in Islamic law—and, among the major religions, Islamic law alone—conversion from Islam is apostasy, punishable by death. Once a Muslim, always a Muslim; attempt to renounce the faith, and both the convert and he who encouraged the conversion are condemned. The Hadith is quite specific on this point: "Whoever changes his religion, kill him" (Hadith, 37). While apostasy and blasphemy (another capital offense, as Khomeini's fatwa against British author Salmon Rushdie reminded the world) may not routinely result in execution, it is noteworthy that reservations against the principle are almost never heard from clerics or commoners, and opposition to condemnations is rare. In Iran in 2002, when a conservative court sentenced a university professor to death for publishing a proposal for an Islamic "enlightenment," thousands of university students did take to the streets in protest. That such a sentence could be handed down at all is indicative of the gulf between the dogmatic and the rational in this historic and civilized society. It is also indicative of the power and prevalence of the dogmatic and the paucity and scarcity of the rational. Thus to express revulsion, or even disapproval, of acts of terror perpetrated in the name of Islam entails risks no "moderate" Muslims can afford to take and few mullahs or imams would encourage.

Importantly, Muslim rage is not directed only at the West in general or America in particular. Terrible, death-dealing conflicts have marked the Islamic realm as they have the Christian world, conflicts born of sectarian issues as well as other, more worldly causes. The war between Iran and Iraq in the 1980s, which may have cost as many as a million lives, pitted Shi'ite Iran against a Sunni-ruled Iraq, but it was not primarily a sectarian conflict. The invasion of Kuwait by Iraq in 1990 was over egress and oil, not religion. The strife that began in western Sudan in 2003 and ravaged the Darfur Province on the border with Chad was over land, not faith. Acts of terrorism directed against rulers (such as the 1981 assassination of President Sadat of Egypt) and others reflect the volatility of Islamic society. "In the early 1990s Muslims were engaged in more intergroup violence than were non-Muslims, and two-thirds to three-quarters of intercivilizational wars were between Muslims and non-Muslims. Islam's borders are bloody, and so are its innards" (Huntington, 1996). Ten years later, the escalation of Islamic terrorism has confirmed both observations.

Islam, as a belief system, is six centuries younger than Christianity. It is worth considering where Christianity was in the 1400s, the time when Joan of Arc was burned at the stake after a church court condemned her of heresy. Roman Catholic ecclesiastical abuses were leading to the Protestant Reformation, and Martin Luther would soon enter the stage. The Spanish Inquisition was authorized by Pope Sixtus IV to combat apostate Jews and Muslims and to pursue heretics; thousands were burned at the stake and countless more were tortured and dispossessed. Still to come was the terrible strife between Catholic and Protestant forces with incalculable civilian casualties and immeasurable cruelty; Catholic armies could have taken a page from the Quran when it came to dealing with unbelievers. In their Bible, Christians found encouragement in the chapters of Deuteronomy (and elsewhere) to engage in unspeakable barbarity. The Enlightenment still was 300 years away.

In the centuries that followed the Reformation, Christians learned to accommodate their sectarian differences, to live in reasonable harmony with each other, to constitutionally separate church and state, to guarantee citizens not only freedom of religion but also from religion. Except for Northern Ireland, where the conflict has involved more than religion alone, Catholics and Protestants are not killing each other over their religious preferences. Europe has been described as being in a "post-Christian" stage as churches are emptying and congregations are ageing, but this trend is a matter of free choice and there is no coercion to reverse it. Christian fundamentalism has expanded on the religious and political maps of the United States, but it has no violent dimension. Some observers suggest that Islam needs and awaits a Reformation of its own, and the Muslim world an Enlightenment that will obviate the frustrations and anger now pervading the culture. But Islam is not organized the way hierarchical Christianity is, with popes and bishops whose decrees can control the most remote congregations. While there are ayatollahs and imams in the more structured Shi'ite sect, the overwhelming majority of Muslims are Sunnis, and Sunni clerics are masters of their mosques. No edict or decree will change this; it has taken the Saudi political regime, not the religious establishment, to reign in some (by no means all) of the most extreme voices from the pulpits.

So the disheartening prospect is of increasing Islamic violence against Western targets, eventually entailing the risk of the use of weapons of mass destruction. The discussion of China referred to the high-risk potential for an intercultural Cold War to follow the intracultural Cold War that pitted the United States against the Soviet Union. This leads to the parallel possibility that the intrareligious struggles *within* Christianity that yielded today's accommodations will be followed by an interreligious confrontation *between* the only two truly global religions with the potential to resemble the earlier conflict, but at an even more destructive level given the population numbers at risk and weaponry within reach. Importantly, this looming conflict is not a two-way struggle for the minds of those involved. Apart from a few misguided evangelical proselytizers, Westerners are no longer seeking to convert the rest of the world to Christianity. It is Islam's aggressive believers who are sure that their actions on behalf of Allah are a sacred duty, and that their holy war is just, no matter what the cost.

MIGRANTS AND MILITANTS

Go to a Muslim country as a visitor, and you are likely to be treated with memorable kindness and courtesy by ordinary citizens whose hospitality is legendary. But become one of the tens of thousands of Westerners who have become temporary but long-term residents in oil-exporting countries from Algeria to Saudi Arabia, and you are likely to find your hosts less receptive. Public opinion toward foreign residents in Saudi Arabia has changed quite drastically from the days prior to the Gulf War. The presence of United States armed forces and bases in the land of Mecca was anathema to many, leading Washington to withdraw most of its troops—but not before a series of terrorist attacks on foreign military and civilian residential areas caused numerous casualties. Not only the military but many family members and nonessential employees have left Saudi Arabia. On this front in the War on Terrorism, the terrorists and their allies have had some success.

Their ultimate goal, of course, is to rid Saudi Arabia (and the entire Muslim realm) of all Westerners, to erase the symbols of Western imperialism and intervention, and thus to protect the culture against undesirable influences. The American invasion of Iraq made their worst dreams come true: while there had been some public support for the ouster of Saddam's forces from Kuwait in 1991, the Saudi regime could not even afford to allow the use of United States bases as launch facilities in 2003, so strong was domestic opposition to the initiative. When the Iraq campaign devolved from conquest to insurgency, money flowed in large quantities from Saudi sources to Iraqi insurgents (Schmitt and Shanker, 2004). But this will not satisfy the militants who have Saudi Arabia in their sights. Overthrowing the monarchy, ousting the Westerners, closing the oil industry, and establishing a theocracy will ensure that Saudi Arabia does not meet the fate of Iraq.

The number of Westerners resident for economic, diplomatic, or other purposes in the Muslim world, however, is dwarfed by the millions of Muslims who are streaming into the West as immigrants. The dimensions of this migration stream are unclear, because part of it is illegal and unrecorded. The number of adherents to Islam in the United States is variously reported to lie between 4 and 6 million, but a relatively small percentage of this figure represents recent immigration. Long before immigration from the Islamic realm gained momentum, America had its own Nation of Islam, a movement based in the African-American community that began in the 1910s, gained momentum in the 1930s, and had a membership of about 0.5 million by 1975. In the mid-1980s, then called the American Muslim Mission, the organization disbanded although its mosques kept functioning, and the original name was assumed by a splinter group based in New York led by Louis Farrakhan. Today, the biggest concentrations of Muslim immigrants in America are clustered in eastern Michigan, notably the Flint and Detroit areas, upstate New York, including the Buffalo area, and Los Angeles, where many Iranians have settled. Even if the higher figure among estimates is accurate, however, adherents to Islam, domestic and foreign, account for just 2 percent of the United States population.

A proportionately far larger Muslim immigration stream has arrived in Europe, and continues to do so. Europe, too, had clusters of Islamic population long before this latest invasion began: the Ottoman Empire left behind millions of converts in Eastern Europe whose descendants now make up 80 percent of the UN-administered territory of Kosovo, 70 percent in Albania, 43 percent in Bosnia, and 30 percent in Macedonia (Fig. 8-3). But the total in these small countries is dwarfed by the millions who have recently come to France, Germany, the United Kingdom, and Spain. In 2005, Europe as a geographic realm had an estimated Muslim population of about 30 million, more than 5 percent of its total, and Western Europe alone had 15 million (by comparison, the numbers of Westerners in Muslim countries are counted in the tens of thousands).

Europe's Muslim population has diverse sources. The majority of Germany's approximately 4 million Muslims are Kurds from Turkey. Many of the United Kingdom's more than 2 million Muslims come from India, Pakistan, Bangladesh, and North and West Africa as well as Southwest Asia. The leading source of France's nearly 5 million Muslim immigrants is Algeria. Spain has taken in nearly 1 million Muslim immigrants, mostly from Morocco, many of whom have entered the country illegally across the Strait of Gibraltar.

The majority of these immigrants are politically aware and culturally insular. They continue to arrive in a Europe where native populations are stagnant or declining, where Christian religious institutions are weakening, where secularism is rising, where political positions and attitudes often appear to be anti-Islamic, and where certain cultural norms are incompatible with Muslim traditions. Many young men are unskilled and uncompetitive in countries where unemployment is already high, get involved in petty

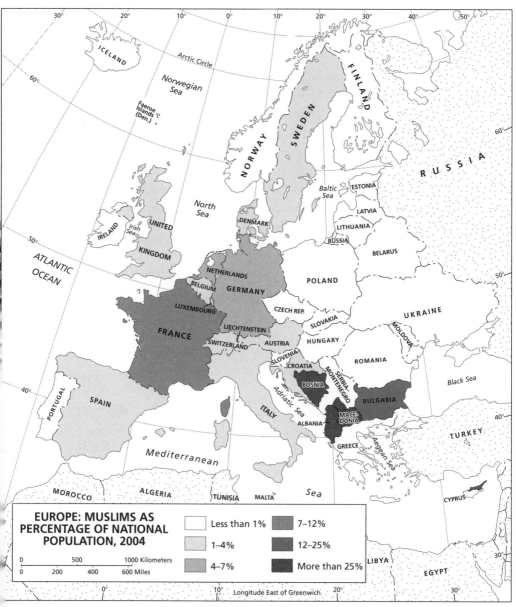

EUROPE: MUSLIMS AS PERCENTAGE OF NATIONAL POPULATION, 2004

Less than 1%	7–12%
1–4%	12–25%
4–7%	More than 25%

0   500   1000 Kilometers
0   200   400   600 Miles

Fig. 8-3

crime or the drug trade, are harassed by law enforcement, and turn to their faith for solace and reassurance. They also sometimes fall under the spell of radical Muslim clerics who find a ready market for their fiery sermons. Unlike most other immigrant groups, Muslim communities tend to resist assimilation, making Islam not only the essence of their identity but also the inspiration for their activism. In Britain alone there are 1,600 mosques, and some of them have become recruitment centers for militants.

The role of European mosques as recruitment centers and militant imams as recruiters are matters of growing concern in the realm. In London, a prominent Muslim cleric known as Sheik Omar attracts hundreds by projecting the collapsing World Trade Center on a giant screen behind him, a spectacle greeted by the faithful with shouts of "Allah Akbar!" (God is Great!) even as the sheik urges young Muslim men to go to Iraq to join the "global jihad" (Van Natta and Bergman, 2005). The role of the al Quds mosque in Hamburg in the formation of the al-Qaeda cell led by the 9/11 terrorist Mohamed Atta is well known, but other major German mosques continue to serve similar purposes. In January 2005, German police arrested 22 Muslims in the mosque-based network of Ansar al-Islam, all actively recruiting militants to fight or carry out suicide attacks in Iraq. French and Italian counterterrorist agencies have traced the paths of numerous militants from their mosques to the scene of conflict.

Restraining the imams and restricting their vituperative speech is difficult under European laws, which tend to protect religious pronouncements—even rhetoric inciting violence. Imams issue calls to war, order members to join al-Qaeda (as Sheik Omar did in one of his London sermons), urge Muslims to murder nonbelievers (the London imam Abu Hamza al-Masri was arrested on this charge), encourage abuse of homosexuals (the court in Rotterdam ruled that such language was within religious bounds), condone physical abuse of women, and command the faithful to resist "assimilation" into the "crusader camp." Concern over what appears to be an intensifying movement has led European states to consider tightening their antiterrorism laws, which engendered intense debates over individual freedoms and human rights as well as angry denunciations and threats of retaliation from Muslims. Surveys indicate that public opinion in Europe supports stronger control over Muslim militancy in its various forms. Europeans in countries with larger Muslim minorities opine that assimilation and integration have failed, and that alternate solutions to the problems of coexistence must be found.

Some European governments did less than others to foster the very integration they see Muslims rejecting. The French, always culturally confident, assumed that their North African immigrants would aspire to assimilation as Muslim children entered public schools; instead, they found themselves in a dispute over the dress codes of Muslim girls, whose head cover was deemed a religious symbol and hence prohibited. The Germans for decades would not award German citizenship to children of immigrant parents born on German soil. The Spanish made it difficult for Muslims to secure permission to build large mosques in some major cities (in 2004, Seville was still resisting such construction). All this contributed to the early radicalization of many young Muslims, some of whom were drawn into terrorist cells.

The presence of large Muslim minorities and their concentration in often-dismal urban housing projects created opportunities for outsiders as well as locals to plot and execute terrorist acts. The expanding network of mosques, many of them built with Saudi, Libyan, and Malaysian funds, created channels for the flow of money as well as militants (Europe's largest mosque, on the outskirts of Madrid, was paid for by the Saudis). In this respect the experience of Spain was telling. The country was shocked when it transpired that the terrorists who bombed commuter trains in early 2004, killing and wounding more than 1,200 people, were Moroccans. This action changed the outcome of the national elections a few days later, when a conservative government that had supported America's war in Iraq was defeated by a socialist who had pledged to withdraw Spain's troops from the country. In order to reduce their dependence on money from Islamic radicals, the new Spanish administration wanted to provide its own financial support for the country's 400 mosques and prayer houses. But this would counter the government's announced plans to achieve greater separation of church and state, so the plan was shelved. The presence of the new Islam in the old Al-Andalus has its ironies.

Western European security and intelligence agencies, keeping track of extremists caught up in the web of Islamic terror cells, found that some Muslim militants had gone to fight in Afghanistan during the anti-Soviet war of the 1980s. Others fought in Eastern Europe during the 1990s, when Muslims were being killed in large numbers by Croats and Serbs during the collapse of the former Yugoslavia. Still others found their way to the Caucasus to join the Chechnyan campaign against the Russians. Virtually all of these militants returned to Europe and joined terrorist cells, becoming a serious and growing problem for law enforcement. European police have thwarted numerous terrorist plans, and they report that almost every detained suspect had been in Afghanistan, Bosnia, or Chechnya. When the United States and its allies invaded Iraq in 2003, a new opportunity to fight in a jihad presented itself, a more attractive one, perhaps, than any of the previous war zones. Press reports in late 2004 indicated that a major Iraq-related recruitment network had been established in virtually all European countries with Muslim minorities. This network was known to provide false documents, money, training opportunities, and transport to places from which infiltration into Iraq would be possible. It was reported that Abu Musab al-Zarqawi, the Jordanian terrorist whose televised beheadings of kidnapped hostages shocked the civilized world, had personally funded and organized a network of recruiters who had already sent about 1,000 European and other militants to Iraq. If these reports are accurate, an end to the Iraq campaign, in whatever form it comes, will present Europe (and other parts of the world) with still another challenge as trained terrorists return to continue their holy war.

Events in the Netherlands cast a troubling shadow over the prospects for Europe. The Netherlands had by some measures been the most liberal and accommodating among those countries receiving Muslim (and other) immigrants, its social policies, including those relating to drugs that are illegal elsewhere, creating an atmosphere of freedom and comfort. By 2004, about 1 million immigrants had settled in the country (population 16 million), and the vast majority had adapted so well to European life that the "Dutch model" was envied across the realm. But the inevitable irritations in a culturally diversifying society, coupled with concerns over the rising costs associated with immigration generally, led to a public backlash that thrust a politician named Pim Fortuyn into national prominence. Fortuyn wanted to restrict further immigration and impose Dutch social norms more stringently on those already in the country, and the popularity of his views was reflected by substantial support among Dutch voters. But in 2002 someone who objected to other aspects of Fortuyn's political position shot him, and his movement soon fell apart. Political assassinations are a rarity in Dutch life—none had taken place for centuries—and the shock was slow to ebb. Then, in November 2004, the filmmaker Theo van Gogh, distantly related to the great painter, was assassinated on an Amsterdam street by a Muslim terrorist who not only shot and stabbed his victim repeatedly, but also left a statement on his body protesting van Gogh's work, which included a film decrying the mistreatment of women in Muslim society. Van Gogh had been a persistent critic of the vituperative pronouncements of Muslim clerics, the erosion of Dutch culture by Muslim immigrants, and the cruelty to animals he associated with Islamic custom. His murder had a major impact on Dutch opinion and was a reminder of the consequences a single act of terrorism can have on the dynamics of society—a point undoubtedly not lost on those returning to Europe with the experience gained in Iraq and elsewhere.

Is there any positive dimension to these generally grim developments? It is often noted that another part of the world with democratic traditions and a (much larger) Muslim minority has managed, despite continuing spasms of violence and assassination, to bring that minority into the democratic process, even as nationalism surged. That part of the world is India, whose population, according to its 2001 Census, is nearly 14 percent Muslim, representing the world's largest cultural minority (140 million) by far. One reason for India's remarkable success lies in the high percentage of younger people in its fast-growing Muslim minority, young adults who may live in Muslim communities but who see the opportunities in India's burgeoning economy and who are adapting their religious life to the practical realities of the present. Strife between Hindus and Muslims does occur in India, sometimes over religious flashpoints such as the temple/mosque at Ayodhya, and occasionally with loss of life. In violence-afflicted Kashmir, no final accommodation is in sight. Still, considering the dimensions of the country's Muslim population sector (in addition to the Sikh and numerous other, smaller cultural minorities), and the nationwide distribution of Muslim communities, which are present in every one of India's 28 States, the modern record of religious coexistence in India is encouraging.

Some cultural geographers and other scholars have drawn parallels to Europe where, they anticipate, a new Islam may ultimately evolve among the children of immigrants who will formulate a modern adaptation of their faith in which pragmatism rather than fundamentalism prevails (Kepel, 2004). They note that Muslim immigrants in Europe attend mosques in numbers proportionately lower than those in their countries of origin, and argue that the size and numbers of new European mosques reflect recent surges in immigration rather than a growing commitment to the faith in the Islamic communities already present. The number of mosques in the militant network is small compared to the thousands of mainstream mosques across the continent, suggesting, to these observers, that the market for extremist activism remains small. Transculturation, then, could yield a more liberal European Islam that may even diffuse into, and modify, the tradition-bound Islamic heartland itself. The question is, in this fast-changing world, whether there will be time for such a slow process to succeed.

9

FROM TERRORISM TO INSURGENCY

Americans since 9/11, and Europeans and Russians much earlier, have learned to live with a sense of unease arising from the randomness and horror associated with terrorist attacks. It is a bit reminiscent of getting on an airplane in the 1950s and 1960s, when airliners, in America and Europe at least, crashed in far greater numbers than they do today, and the statistical chances of a passenger becoming a fatality were much greater. Television coverage and front-page press reports of such accidents made you aware that you were taking a chance when booking a flight. This did not change travel habits, but it was always in the back of your mind.

Today we live with color-coded alerts (whoever thought this system up has a lot to answer for) and constant warnings about terrorist plans to act during national and religious holidays, year-end celebrations, major sporting events, or other prominent occasions. After being told of "specific" threats we are told to go about our business as usual. Airplane passengers, not security guards, overpowered the only shoe-bomber to get on an aircraft. The airport inspection fiasco (remember those "random" gateside searches?) has damaged the industry and discouraged air travel in ways those accidents of the sixties never did. Until terrorists acquire and manage to use weapons of mass destruction, an individual's chances of dying in a terrorist attack are less than they are of dying in an automobile accident.

On the other hand, certain regions and places are at greater risk than others in this era of terrorist threats. The 1993 attack on New York's World Trade Center was a failure by al-Qaeda standards, since the acknowledged notion was to topple one or both of the towers through a massive explosion at their base. That attack may or may not have been planned by al-Qaeda operatives, but it should not have come as a surprise that another effort was made in its wake. The failure of the second 9/11 attack on buildings in the Washington area because of the heroism of passengers aboard United 93 is likely to be a factor in al-Qaeda priorities in the years ahead. Prominent landmarks, historic edifices, and crowded facilities are obvious targets because their destruction will refresh the emotional impact of acts of terror on the American public. Fear-inducing publicity is a primary goal of terrorist actions.

Such factors obviously play a role in the selection of targets and their location. Terrorist organizations calculate the psychological impact of their attacks as well as the damage inflicted. But their most successful strikes also alter the behavior of entire societies with major economic implications. The United Nations Panel on Threats, Challenges and Change estimated in their 2004 report to Secretary General Kofi Annan that the 9/11 attacks, in addition to their direct costs in lives and property, cost a further $80 billion in national and international economic losses and pushed 11 million people in developing countries into poverty. The economic fallout from an attack involving weapons of mass destruction would dwarf these numbers (Annan, 2004).

Part of this cost represents the impact on the airline industry, from the grounding of all American aircraft on the day of the attack and the shutdown of air transportation for several days afterward to the temporary but crucial fear-induced shrinkage in passenger volume. The chaotic and inconsistent installation of airport security systems further damaged the industry. Today the system works better. But a RAND Corporation study published in January 2005 raises the prospect of potential attacks on commercial aircraft in the United States by terrorist operatives using shoulder-fired missiles because hijacking an airliner by boarding it is now less feasible. A key segment of this report, however, raises a troubling cost assessment: about $1 billion for the aircraft and its passengers and crew, $3 billion if the aviation system were again shut down for up to a week, and another $12 billion resulting from lost business and reduced passenger loads following an attack. Elsewhere, the report states that it would cost about $11 billion to equip the nearly 7,000 commercial jets based in America with missile defense systems, and that annual maintenance of these systems would require more than $2 billion. Since the cost of a single attack would exceed this investment, it would appear that it is worth making.

Terrorist organizations are well aware of such calculations when they plot attacks, although the disruption caused by the 9/11 attacks probably exceeded their expectations. The prospect of again grounding the entire American airliner fleet for a week following a single successful missile attack may form a stronger motivation than the success of the attack itself. Signaling that such an event would be treated as would an accident, with service continuing without interruption, could constitute a strong deterrent and a counterstrike in the psychological contest that attends terrorist actions.

A rapid return to normalcy is one way Israel responds to the frequent suicide attacks it has faced. The scale of these attacks should be seen in the context of the country's population, which is one-fortieth of that of the United States. When an attack kills 25 people on a bus or in a restaurant, the equivalent in America would be 1,000 casualties. Yet Israel, very early in its experience with terrorist actions, embarked on a policy to clear the attack scene immediately and return it to near normalcy in a matter of hours, denying the terrorist network the satisfaction of achieving long-term disruption as well as carnage. It is a practice that should be part of American preparation for terrorist attacks that may or may not materialize.

ROUTES AND REGIONS OF RISK

Several global and regional Islamic terrorist organizations are currently active, and their coordination improves as their reach expands. European Muslims under the aegis of Ansar-al-Islam as well as al-Qaeda are currently participating in the insurgency in Iraq, and terrorist cells are known to exist around the world. Iraq, according to pronouncements made by leaders of these organizations, is a training ground for the conversion of isolated attack to coordinated campaign.

Where do Islamic terrorists prosper? Certain geographic circumstances and environments are especially favorable to them. First, failed or malfunctioning states provide a unique opportunity for infiltration because their disorder permits terrorist cells to organize and operate beyond the control of authorities, or even with their connivance as was the case in Afghanistan. Today, Somalia is another failed and significantly imperiled state, its territory divided into three parts of which the southernmost sector, centered on the chaotic capital of Mogadishu and adjacent to vulnerable Kenya, is a breeding ground for terrorism. In the Western Hemisphere, chronically unstable Haiti presents an opportunity for terrorists seeking to establish a base. The number of malfunctioning states elsewhere is increasing. Several, such as Pakistan, Turkmenistan, Eritrea, and Côte d'Ivoire, lie on the margins of the Islamic realm. Others, such as Paraguay, Venezuela, and Cuba, raise concerns in the United States because of their relative location in the Americas and their potential role as bases for links between terrorist headquarters and operatives in the field.

Second, terrorists like large, chaotic cities because they provide cover and anonymity as well as amenities such as phones, banks, mosques, and money-laundering businesses. Manila and Jakarta in Southeast Asia, Karachi and Peshawar in South Asia, Cairo and Casablanca in North Africa, Mombasa and Cape Town in Subsaharan Africa, São Paulo and Guayaquil in South

America are among cities that have harbored terrorists and undoubtedly continue to do so. Of course, so did New York, Los Angeles, and other American cities during the planning of the 9/11 attacks, but it was intelligence failure, not the protective urban environment, that led to their success. Intercepting intercity phone messages and tracing interurban money transfers are among the most effective counterterrorist strategies, but the larger and less formal the city, the more difficult the task. Excellent police work in London, Paris, and Hamburg—to name just three European cities where numerous terrorist plans have been uncovered and the plotters arrested—is helped by the comparatively orderly urban environments prevailing there.

Third, terrorists like remote, rugged, rural environs where they can blend in with the local population while operating their networks or preparing their attacks, especially when those environs adjoin suitable targets. Of course the mountainous, forested eastern periphery where Afghanistan adjoins the Tribal Areas (also called Waziristan) of Pakistan is the prime example, the place where both Usama bin Laden and his Taliban ally Mullah Omar sought and found refuge and from where bin Laden's second in command, Zawahiri, continues to direct operations. But no one should have been surprised at the deadly attack on the Bali nightclub in 2002 that killed about 200 people, most of them vacationing Westerners. The populous Indonesian island of Jawa (Java) has two principal refuges for Islamic terrorists: the huge and turbulent city of Jakarta in the west and the extremist-infested east, where the country's main fundamentalist movement also has its base. The Hindu-cultured, tourist-dependent island of Bali lies just across the water from this hotbed of Jemaah Islamiyah (JI) whose local leader, Abu Bakar Bashir, was put on trial for the attack. But because the court determined that JI did not officially exist, and despite evidence provided by the arrested bombers that Bashir was the head of the organization and had ordered the assault, he was convicted only on treason and immigration charges and sentenced to four years in prison, reduced on appeal to three. Such a setback illustrates how difficult "winning" the War on Terrorism can be.

Given these geographic advantages, sought by terrorist leaders and exploited by their operatives, what conclusions can be drawn? Let us look at two cases, one local, the other regional. The juxtaposition of globally connected and despised Western targets and remote areas of Islamic militancy is not confined to Bali. On the peninsula where Thailand meets Malaysia is a similar situation. Thailand's southern provinces have become a hotbed of Islamic militancy, in part because of the porous border between the two countries (the Indonesian terrorist known as Hambali, arrested in Bangkok, had been seen in the area; he had apparently crossed the border, which is marked by a river, from Malaysia without documents). Most of the Islamic militancy, which has included the murder of Thai police, has been concentrated on the east side of the hilly peninsula. But just across, on the west side, lies the internationally famous (now tsunami-damaged) resort of Phuket, playground of wealthy Westerners and Asians. The risks are obvious. Another such juxtaposition occurs in and off the southeast Philippines, scene of a long-term Muslim rebellion against Manila's government that has become enmeshed in full-fledged terrorism, most of it perpetrated by the movement known as Abu Sayyaf (the United States recognizes three other organizations in the area as terrorist movements). Less than 50 miles (80 km) away from the Jolo archipelago, where Abu Sayyaf is based, lie luxurious Eastern Malaysian waterside resorts such as Sipadan. From there, these terrorists kidnapped Western tourists in 2000, with deadly consequences. There are other, similar cases. While there is no reason not to travel at all to remote places in the world, be aware of developments and read the map before you go.

RISK IN THE WESTERN HEMISPHERE

Changing our scale from the local to the regional, consider the implications for United States security of developments in South and Middle America over the past few years. Until 9/11, the most serious Islamic terrorist attacks in the Western Hemisphere had been the 1992 bombing of the Israeli embassy in Buenos Aires and the 1994 bombing of the Jewish Community Center on the city's fashionable Florida Avenue, resulting in more than 120 deaths and hundreds of wounded (the domestic-terror truck bombing of the Federal Building in Oklahoma City in 1995 was the worst overall). Almost immediately after the 1992 attack, suspicion fell upon Lebanon-based Hizbullah in collusion with staff members of the Iranian embassy in Buenos Aires, but no concrete evidence could be secured because of diplomatic immunity. Argentina, with nearly 1 million practicing Muslims, has a long historic link with Lebanon; about 20 percent of the country's population has Arab ancestry and its former president, Carlos Menem, is among them.

Following the 1994 Jewish Community Center attack, pressure to investigate and solve the matter heightened, but high-level interference slowed the process down. In 1998, Argentinian judicial and intelligence officials got a break when an Iranian defector in formal testimony implicated senior members of the Tehran government including the president, the minister of foreign affairs, the head of intelligence, the son of the ayatollah Khomeini and the Iranian ambassador to Argentina at the time of the attacks. Additional information, gathered by Argentinian intelligence, showed Iranian officials leaving and entering Argentina under false names around the time of the bombings (Rother, 2002a). For the first time in this investigation, a geographic connection was made that continues to matter today: the so-called Triple Frontier where Argentina, Brazil, and Paraguay converge (Fig. 9-1). Telephone calls from the Iranian embassy to Hizbullah members living in this remote area revealed the organizational link between the Iranians and this terrorist network, and the location of its South American base.

The Triple Frontier, referred to as the Triborder Area by the United States State Department, has an Arab immigrant population exceeding 25,000 and is described as "teeming with Islamic extremists and their sympathizers, [where] businesses have raised or laundered $50 million in recent years" (Rother, 2002b). A map of the area was found in an al-Qaeda Kabul safe house in 2001, raising concerns that al-Qaeda might have established a base here. Press reports in 2002 told of two Lebanese men who had been arrested in Paraguay raising money for the organization, but specifics have been scarce.

What is certain is that the Triple Frontier satisfies the criteria just discussed: a huge, chaotic city not far away (São Paulo with its international and continental linkages, its large Muslim population, and its vast expanses of poor *favelas*); a quiet and remote rural area with villages where Muslim merchants can establish themselves and carry on supportive activities; a malfunctioning state in the form of Paraguay that does not, for example, have an antiterrorism law; and Western targets near and far for attack. As Rother reports, the Triple Frontier has become a haven for militants, fugitives, forgers of passports and credit cards, telephone switching operators, money launderers, and enforcers. In the process, the cultural landscape in this area has drastically changed. In the neat, trimmed town of Foz do Iguaço, the tourist stopover on the way to the nearby Iguazu Falls, the television stations available in a visitor's hotel room include Brazilian stations and CNN International but also five Arab-language stations, two of which broadcast round-the-clock sermons from Lebanon, Saudi Arabia, and elsewhere in the Muslim world. Near the town's peak-value intersection stands a gold-domed, finely crafted mosque, and in the streets numerous women wear the burqa. Take the bridge across the Parana River to the Paraguayan side, and the town of Ciudad del Este is a vivid contrast to neat and orderly Foz—a jumble of mostly low-slung buildings in poor repair along traffic-clogged streets with broken sidewalks, a town where contraband smuggling, a busy drug trade, illegal arms sales, the manufacture and marketing of pirated goods, prostitution, and petty crime are endemic. But locals say that there are more than 40 mosques and prayer houses here, and that Islamic extremists are recruiting and indoctrinating followers.

Which raises the question whether the Triple Frontier is a staging area for a wider campaign. The Argentinian press reports that known militants' telephone traffic, fiscal transactions, and travel patterns indicate links not only to São Paulo and Lebanon but also to disorderly Guayaquil, one of South America's toughest cities, and to sprawling Maracaibo in Venezuela. Where,

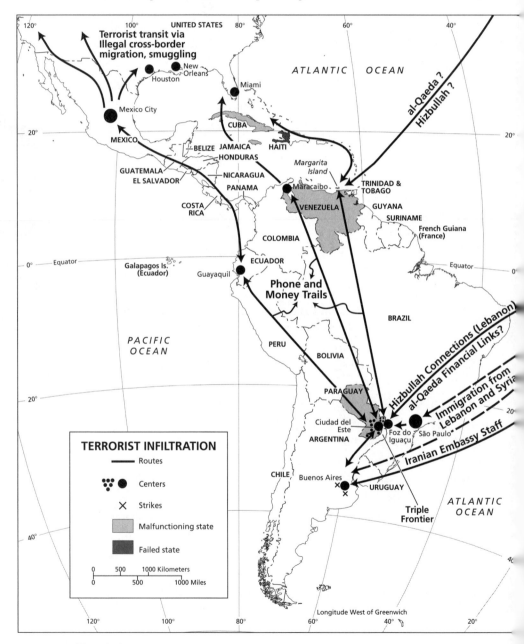

Fig. 9-1

beyond these cities, the trails lead is uncertain, but consider this: the United States is flanked to the north by a dependable neighbor and to the east and west by wide oceans, but to the south it is exposed to access in various forms, from illegal immigration overland to stepping-stone entry via islands in the Caribbean. The logical route for terrorists would surely be from the south, and the staging area may very well be the Triple Frontier.

In recent years the situation has become still more complicated because Venezuela has taken on the characteristics of a malfunctioning state. While the situation in neighboring Colombia, also malfunctioning and under terrorist threat, has remained a largely domestic matter, President Chavez has taken Venezuela in a new direction, proclaiming ideological affinity with two terrorist groups (FARC and ELN) in Colombia and allowing Venezuelan weapons to reach them across their joint border (U.S. Department of State, 2004). President Chavez also proclaimed solidarity with Cuba's dictator Fidel Castro, but the more alarming effect of Venezuela's political condition is the reported appearance of radical Islamic operatives in the country, notably on Margarita Island, part of the northeastern Province of Sucre and on the doorstep of the Caribbean (Fig. 9-1). A malfunctioning political system in this pivotal location is an opportunity sure to be seized by Islamic militants.

As infiltration into the United States by land, air, and sea from the north, east, and west becomes more difficult because of increasingly effective controls, the map suggests that southern routes prominently including those via the Mexico-U.S. border, will become crucial to terrorist operatives in the years ahead, and that potential southern targets will require special homeland-security attention. From Caribbean shipping lanes to Gulf of Mexico oil installations, and from Houston to New Orleans and Miami, America's southern flank requires protection against designs that may be on the drawing boards elsewhere in the world and transmitted to the Western Hemisphere via the Triple Frontier.

THE ISLAMIC FRONT IN AFRICA

Islam spread along the northern shores of Africa by caravan and along the eastern coast by boat, establishing itself in modern-day Egypt before the end of the seventh century and reaching Morocco before the end of the eighth. The faith's rapid diffusion transformed not only what we call the Middle East today but also North and East Africa, reaching Dar es Salaam, Tanzania by the eleventh century—about the same time it penetrated modern China after sweeping across Central Asia. From their bridgehead in North Africa, the Moors (or *Mors*, hence *Morocco*) invaded Iberia, where flourishing Islam left indelible imprints on the cultural landscape.

While ruling al-Andalus, the Muslims also looked southward, to West Africa across the Sahara. There, on the savannas between the forests of the coast and the desert of the interior, lay a tier of thriving African states in control of trade and traffic. From what is today Senegal in the west to northern Nigeria in the east, these states benefited from what economic geographers call "double complementarity": the peoples of the steppes to the north needed goods available from the forests of the south (starches, spices, animal products, building materials) while those of the forests wanted items from the interior such as leather and salt. These goods were traded on the bustling markets of the savanna states, and the Niger River ("where the camel met the canoe") was the Mississippi of West Africa. Through these markets, attracting the attention of the Muslims of Morocco, moved a considerable quantity of gold and gemstones.

Stable, durable, and in control of large populations and territories, the West African states such as Ghana, Mali, Songhai, and others were tempting targets for the powerful Muslims of the north. Caravans carried not only Moroccan leather but also Muslim proselytizers, and before long the kings and chiefs of West Africa's savanna states were converted to Islam and were ordering their subjects to follow suit. This led to a momentous cultural reorientation, because now the savanna corridor between forest and desert stretching across Africa from west to east became a conduit to Mecca. Ghana fell apart when the Muslim invasion was backed up by armed force, but Mali survived, and from there, gold-laden annual pilgrimages involving tens of thousands of the newly faithful walked eastward, making Khartoum a gathering point before heading for their final stage to Mecca. Khartoum became a West African outpost, its population swelled by pilgrims who never made it to Mecca and others who stayed there on the way back.

Christianity, with its six-century head start, did not make an impact in West Africa until colonial times, but in Africa's "Horn," where Ethiopia lies today, several African states had accepted Christian beliefs: Kush, Nubia, and Axum among them, the last giving rise to the Christian dynasty that eventually shaped modern Ethiopia. When Islam reached this area, it could not overpower these kingdoms, and Christianity survives there to this day, virtually encircled by Islamic societies.

Islam's southward march was halted by the arrival of European colonists, who established beachheads along the West and East African coast and moved inland, taking control not only of peoples who practiced traditional African religions but also those who had adopted Islam. The modern map of

West Africa bears witness to this episode: boundaries between former British, French, and German possessions run from the coast into the interior, sometimes for dozens, elsewhere for hundreds of miles. Britain's Nigeria extended from the coastal forests to the interior desert, and its girdle of boundaries threw together Muslims and non-Muslims alike. Then the colonists set about promoting Christianity among their subjects, so that Nigeria today has a Muslim north and a mainly Christian south (traditional African religions do survive in Africa's "Christianized" areas), creating severe tensions that lead to recurrent riots and political strains that could eventually endanger the state.

When the colonial powers gathered in Berlin in 1884 to finalize their partition of Africa, they paid little heed to the religious divide that stretched across the continent from Guinea in West Africa to Kenya in East Africa. Not only Nigeria but other West African states, as well as Chad, Sudan, and Ethiopia to the east, found themselves with regional-cultural contrasts fraught with problems. In the process, the colonial powers created a regional problem whose consequences they could not foresee. The Islamic Front has become a zone of conflict that threatens the cohesion of countries and facilitates the actions of terrorists and insurgents.

Undoubtedly the costliest impact of the divisive Islamic Front has been in Sudan, whose Muslim government waged a decades-long, bitter war against the three southern provinces where African Christian and animist communities are in the majority. The cost in terms of casualties and dislocations will never be known; estimates of the death toll over more than three decades of conflict are in the hundreds of thousands. The fundamentalist-dominated Sudan regime wanted to impose Islamic (Sharia) law in the South; Southerners wanted independence or, failing that, protection and guarantees against Khartoum's cultural domination. Finally, in 2005, a settlement of sorts was reached, but almost simultaneously another deadly conflict broke out elsewhere near the Islamic Front, this time in Sudan's Darfur Province along the border with Chad, costing tens of thousands of lives even as the United Nations deliberated over the question of whether what was happening there could be technically construed as genocide.

In West Africa, the Islamic Front's proximity to Côte d'Ivoire proved costly in 2004 as well. Long a stable and relatively prosperous country, Ivory Coast had had a northern Muslim minority ever since its territory was bounded. But in postcolonial years, Muslim herders and farmers from neighboring Burkina Faso to the north have been crossing the border and settling in rural areas of the country. Initially there were occasional skirmishes over crops damaged by Muslim pastoralists' cattle and land taken by immigrants, but then the growing strength of the Muslim minority began to

translate into political power. When the country's long-term ruler died and a
Muslim politician sought the president's office, he was disqualified by southerners and civil war broke out, ruining the cocoa-based economy and precipitating French intervention. The French force was accused of favoring the
Muslim northerners, the large French expatriate community was set upon
by mobs, and even the modern, once-thriving port of Abidjan was engulfed
by violence. The effects of the Islamic Front had reached the sea.

If the breakdown of order in Côte d'Ivoire was a shock to observers of the
African scene, simultaneous events in Liberia were even more surprising.
Liberia has long been in the grip of ethnic conflict, but in late 2004 reports
from the capital, Monrovia, described serious clashes between Muslims and
Christians in several parts of the country. This, as far as I have been able to
determine, is the first instance of such religious strife in the country's history. As Figure 9-2 shows, Liberia lies to the south of the Islamic Front, but
the Islamic component of its population has been rising and is now 16 percent. Still, Liberia's Muslim minority is scattered in the main towns, not in
the northern region of the country. Whether the Muslim-Christian clashes
are an isolated event or a signal of more serious problems is not yet clear.

As Figure 9-2 shows, the Islamic Front loops into the Horn of Africa, co-
inciding very roughly with the border between mainly Muslim Eritrea and
dominantly Christian Ethiopia before dividing the latter into a Muslim east
(Ogaden) and a Christian west. The Muslim east is the historic home of the
Somali people living on the Ethiopian side of the border with the failed state
of Somalia that, as the map indicates, is virtually 100 percent Muslim. Ethiopia's Christianized core area, centered on the capital of Adis Abeba, lies in
the highlands, a natural fortress that has protected the country in the past
and from where the founding emperor, Menelik, extended its power over
the encircling plains. In the process Ethiopia's Christian rulers gained control over the Ogaden with its Muslim Somali population and its leading city,
Harer (Fig. 9-2). Thus the Somali found themselves on both sides of a superimposed border, but after the colonists withdrew it mattered little. Clan
allegiances and divisions, not invented boundaries, dominated life. Somali
pastoralists drove their herds across the borders in pursuit of seasonal rains
as they had for centuries, and in Somalia they fought among themselves over
influence and dominance.

The approximate position of the Islamic Front in Ethiopia is of less concern that its extension along the Somali-Kenya border, because Somalia has
been a failed state for decades and it is there, rather than in Muslim Ethiopia,
that the greatest potential for terrorist activity exists. In early 2005, Somalia
remained a state divided into three parts; Somaliland, the former British

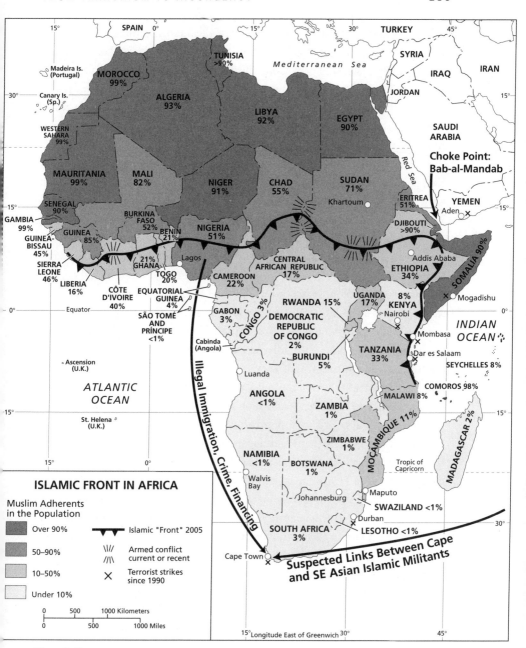

Fig. 9-2

colony in the north, doing best in every way; Puntland in the middle, seeking to distance itself from the rest, and rump Somalia in the south, centered on the capital and largest urban agglomeration, Mogadishu. It is the southern third of Somalia that is in the worst condition, in effect a territory without the rule of law.

Long before Somalia became linked with fears of Islamic terrorism, Kenya had been experiencing the effects of terrorist activity based on the Muslim side of the Islamic Front. Somalia *shifta*, marauders on the move, have been attacking targets in northeastern Kenya ranging from game lodges to archeological digs and from cattle owners to bus drivers for many years. They have struck as far west as Lake Turkana, disappearing into the Ogaden on their way back to Somalia; they have traded their sharp-edged *pangas* for guns, and they are a menace. But they do not have an ideological agenda: that dimension only appeared during the 1990s, when East Africa became the stage for al-Qaeda's attacks on United States embassies in Nairobi and Dar es Salaam and subsequent assaults on a tourist hotel near coastal Mombasa and an Israeli airliner departing Mombasa Airport. As in the case of Bali, note the juxtaposition of refuge and target: Western interests in East Africa's cities lie short distances from areas as tribal as Pakistan's Waziristan.

Is the Islamic Front static or mobile? The evidence from Côte d'Ivoire to Kenya suggests the latter. But even as the Islamic Front moves southward, African countries well beyond its leading edge form obvious opportunities for Islamic expansion. The horrific experiences of Rwandans, Congolese, and the citizens of other malfunctioning African states (or parts of states) may be so strongly associated with Christian contexts—many Rwandans were herded into churches and killed with the connivance of ministers—that Islam may appear as a hopeful alternative. Although data may be unreliable, it is noteworthy that Rwanda two decades ago reported that 2 percent of its population was Muslim; today it is reporting 15 percent, a huge increase in a Roman Catholic country.

Figure 9-2 shows the percentages of Muslims in the populations of African countries, and the Islamic Front effect is obvious from these data. In general, the closer to the Islamic Front, the higher the Muslim component. Tanzania's high percentage, much higher than Kenya's, results in part from its merger with Zanzibar, since almost all Zanzibari are Muslims. But even so, Tanzania's coastal zone, including Dar es Salaam, is strongly Muslim, and may become a regional extension of the Islamic Front. Note that countries in approximately the same latitude—the Democratic Republic of Congo, Malawi, Zambia, Angola—report a minimal Muslim presence.

Which leaves the country farthest from the Islamic Front but prominent in Islamic history: South Africa. More than 350 years ago, the Dutch began to bring Muslim Indonesians to Cape Town as workers, and these "Malays," as they are often called, built the first prayer houses and established a base for the faith. Over time, these Malays became part of the larger population referred to as the Cape Coloured community, and Islam spread into it: this, during the decades of *apartheid*, was one way to reject the churches that preached oppression. Small, graceful mosques were part of the cultural landscape of Cape Town's historic District Six when it was demolished under apartheid laws to move its nonwhite residents farther from the city center.

Today South Africa's population is about 3 percent Muslim, and some of the country's 1 million Muslims are South Asians, descendants of a later immigration now living in Kwazulu-Natal, in the city of Durban and its hinterland. But Islamic activism, and some incidents of terrorism, center on Cape Town. Soon after South Africa's democratic transition, which unleashed freedoms unheard of under the repressive apartheid regime, an Islamic group began issuing fatwas against pornography, homosexuality, drug use, and other "excesses" of freedom; warnings against commercialism were followed by the bombing of a Planet Hollywood in a new waterfront shopping center. Effective police work rolled up the group responsible, but a low-grade terrorist threat continues. In April 2004 there were unconfirmed reports of a threat and interdicted attack on the visiting cruise ship *Queen Elizabeth II.*

As Figure 9-2 shows, there are no confirmed links between any remaining terrorist cells in the Cape Town area and either Southeast Asian or Middle Eastern networks; the activism of the 1990s appeared to be homegrown. If there is any external link, it may be to Nigeria and have more to do with organized crime than with terrorism. But Cape Town is one of those chaotic, fast-growing cities that harbor terrorists elsewhere in the world. Behind the façade of Table Mountain lie the "Cape Flats," a vast expanse where more than 2 million arrivals have built an informal city far larger than the original. Few contrasts in the world are more vivid than that between the opulence of the Victoria Waterfront and its shiny malls and luxury hotels and the abject poverty of "The Flats" a short drive away. It is another one of those juxtapositions.

FINANCING TERROR

The long-distance routes of travel and transactions shown on Figures 9-1 and 9-2 reflect the fiscal as well as the organizational capacities of terrorist organizations. Mounting terrorist attacks costs money, and shutting off the money supply is a key weapon in the counterterrorist arsenal. Tracing the flow of funds from planners to operatives is a critical indicator of intent and, sometimes, the scale of operations. Moving money requires telephone or Internet contacts, bank transactions, and other negotiations, all of which leave a trail. Mosques, businesses acting as money launderers, fraudulent charities, and other nodes in this international network are points of strength as well as weakness in the Islamic terrorist system.

Where do the millions of dollars circulating in the global Islamic terrorist network come from? Terrorism is a criminal as well as an ideological enterprise, and where illicit money is made—from opium, cocaine, and other drugs, from freelance gold and diamond mining and sales, from smuggling contraband, from theft and extortion—such money will siphon into terrorist networks. It is often said that Usama bin Laden used his large inheritance to enable his cohorts to launch their terrorist attacks, but his personal fortune was dwarfed by his income from other sources. Money collected at mosques in Saudi Arabia and elsewhere, funds from Islamic "charities," and legitimate businesses bin Laden and his allies controlled yielded ample resources to sustain a global campaign of terror; the attacks on New York and Washington are reported to have cost just under a quarter of a million dollars in air tickets to planning meetings, flight-school training, accommodations, and related expenses (Report of the 9/11 Commission, 2004).

Another source of indirect support appears to be coming from certain members of the Saudi royal family who appear, like some Iranians before the fall of the shah, to want to have it both ways. One of Usama bin Laden's stated objectives is the overthrow of the royal regime, but he will, of course, accept money from all sources, even princes who want to be remembered should the time come. Allegations that some of this support goes directly to al-Qaeda may be overdrawn, but Saudi money is going to numerous mosques whose imams are known to espouse extremist causes.

The magnitude of the financial terror-support system was confirmed during the conflicts in Chechnya and Iraq, when a network of recruiters enlisted Islamic fighters in Europe and the Middle East to join the campaigns against the Russians and the Americans. Their numbers run into the thousands; all are provided safe houses, transport to the war zones, and weaponry. They see themselves as warriors, not terrorists, but upon their return to their homelands following their jihad, they will be ready recruits for terrorist activity in the future. Would-be combatants have been arrested on their way through Ukraine, Turkey, and other countries, and some will undoubtedly be stopped on their way back to Europe when their wars abate. But most will be housed in the Muslim neighborhoods of Europe's and the Middle East's big cities and maintained through funds generated from terror-supporting sources. Others will fan out along some of the routes marked on the maps. Wherever they find themselves, they will be ready to resume their holy war in another way. Since 9/11, events have been blurring the line between terrorism and insurgency. Where this will lead is not predictable.

FROM TERROR TO INSURGENCY: IMPLICATIONS FOR IRAQ

The first decade of the twenty-first century is likely to be remembered not only for the escalation of Islamic terrorism but also for the American-led military intervention in Iraq that deposed a ruthless dictator and his regime. The extent to which these two phenomena are linked is likely to remain a matter for debate. In January 2002 in his State of the Union address President George W. Bush identified Iraq, neighboring Iran, and distant North Korea as constituting an "axis of evil" with which the United States would have to deal. With the anti-Taliban campaign in Afghanistan in full swing and UN negotiations regarding Iraq in progress, the prospect of a military attack on any of the three "axis" members seemed remote.

But in fact preparations for an invasion of Iraq were already underway. In his book *Intelligence Matters*, former Florida senator Bob Graham describes a briefing in Afghanistan by General Tommy Franks, following which the general, speaking privately to the senator, stated that "military and intelligence personnel are being redeployed to prepare for action in Iraq . . . what we are doing [in Afghanistan] is a manhunt" (Graham and Nussbaum, 2004).

It is possible to discern seven principal reasons for this diversion from Afghanistan to Iraq. Intelligence reports indicated that Saddam Hussein possessed chemical and biological weapons of mass destruction and might achieve nuclear capability as well. Iraq was seen as an immediate threat to Israel, which had strong allies in the Bush administration. Iraq was cooperating with terrorists in various ways, giving large awards to families of Palestinian suicide bombers. Iraq was defying UN resolutions and was doing business in contravention of UN sanctions. Saddam Hussein and his clique were perpetrating dreadful human-rights abuses on the general population. Iraq's oil industry was in a shambles and could once again become part of the global supply. And, some observers argued, George W. Bush had a personal issue with Saddam Hussein because of the dictator's foiled plot to assassinate his father during a visit to the region following the 1991 Gulf War.

Following difficult and divisive UN negotiations that erased much of the goodwill and support the United States had seen after the 9/11 attacks, American forces supported by British and Australian troops invaded Iraq from the south in March 2003. Turkey had rejected United States requests to use its territory to enter from the north. It was actually believed by some high-ranking members of the American administration that the invading infidel soldiers would be welcomed demonstratively by roadside crowds.

The sequence of events following the initial thrust is all too familiar. Meeting relatively little resistance, American forces reached the Baghdad area in a matter of weeks, entered the city and occupied Saddam's palaces, tore down his numerous statues, disbanded his army and police, witnessed a wave of looting and score settling in the general population, and began the effort to establish a post-Saddam administration. On an aircraft carrier off the coast of California, President Bush declared "mission accomplished."

But the real mission lay ahead, and for this the planners had poorly prepared—so inadequately that one wonders to what extent they were familiar with the historical or the cultural geography of Iraq. In many ways California-size Iraq is the pivotal country in the Middle East, with 60 percent of the region's area and 40 percent of its population (demographic estimates range from 23 to 26 million). With the narrowest of outlets to the Persian Gulf, Iraq has six neighbors including Saudi Arabia, Turkey, and Iran, and significant historic and cultural ties with all of them. The country is heir to the earliest Mesopotamian states that arose in the Tigris-Euphrates Basin, and it is studded with matchless archaeological sites and museum collections, to which disastrous damage was done during and after the invasion from the combat and looting. With its major oil and gas reserves and large areas of irrigable farmland Iraq is also the best endowed of all the region's countries with natural resources (Fig. 9-3).

Iraq's physiography ranges from a mountainous east, where the rugged Zagros Mountains afford relatively few routes into neighboring Iran while the hilly northeast has less relief and greener countrysides, to a flat expanse of desert along the southern and western border with Kuwait, Saudi Arabia, Jordan, and Syria, where population is sparse and overland routes tenuous. Between these two extremes lie the coalescing basins of the country's two great lifelines, the Tigris and Euphrates Rivers with their distributaries and lakes, and here the population concentrates.

Iraq was one of those colonial creations with which European imperialists saddled the Muslim world, centered on historic Baghdad but encompassing peoples and cultures that have strong ties across its borders. About 20 million of Iraq's approximately 25 million citizens are Arabs (the others are Kurds), but the Arab majority is divided on the basis of religion between Shi'ites in the south and Sunnis in the north. Shi'ites outnumber Sunnis by more than two to one, but it is the Sunnis who, through violence and intimidation, have ruled Iraq throughout its modern existence.

Iraqi Shi'ites have had a difficult and complicated relationship with the Shi'ites of the Shia heartland, in neighboring Iran. The map shows part of this story: Arabs actually form the majority in the Iranian province of Khuzestan (capital Ahvaz), the annexation of part of which was one of Saddam's objectives during the war with Iran in the 1980s. What the map does not show is

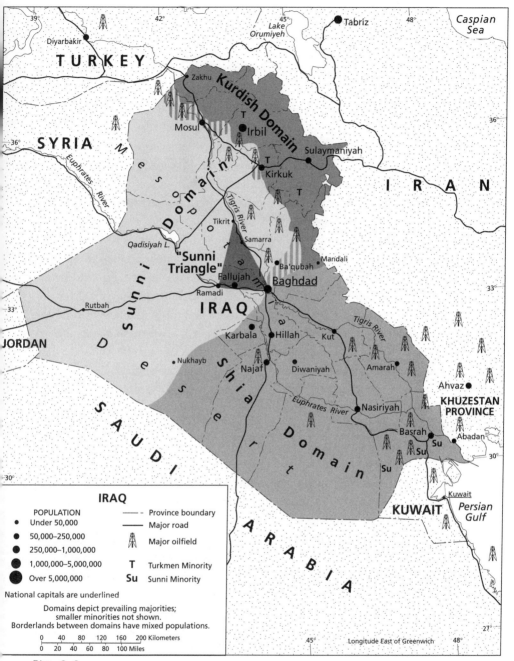

Fig. 9-3

that Arab (Iraqi) Shi'ites and Persian (Iranian) Shi'ites do not see eye to eye. Iraq's Shi'ites adhere to a form of the faith known as Akhbari, which does not have strong political motivation and does not readily generate a political power structure. Iran's Shi'ites follow Usuli doctrines, in which the link between religion and politics is far stronger. As soon as the postinvasion disorder made it possible, Iranian Shi'ites crossed the border and began urging Usuli practices on the Iraqi faithful. One of those who heard that call was a young cleric named Muqtada al-Sadr, who assembled an armed force, holed up in a mosque in the holy city of Najaf, and contributed greatly to the chaos into which postinvasion Iraq descended.

But even Muqtada al-Sadr eventually agreed to participate in the political process, mindful of the order issued by the Shi'ites' grand ayatollah Ali al-Sistani, who proclaimed that failure to do so by any Iraqi is "a betrayal of the nation" that would be punishable by "burning in hell." Indeed, Akhbari doctrine goes far to explain why a comparatively small Sunni minority could mistreat the Shi'ite majority (which constitutes more than 60 percent of the country's entire population) for so long, even to the point of environmental terrorism when Saddam ordered the draining of the historic marshlands of the south, depriving tens of thousands of their traditional livelihoods. Terrible retribution was meted out to Shi'ites suspected of disloyalty following Iraq's defeat in the 1991 Gulf War. Mass graves found after the 2003 invasion contained the remains of hundreds of thousands, most of them Shi'ites. Nevertheless, Iraq's Shi'ites proved themselves repeatedly to be Iraqi Arabs first and Shi'ites second. During the Iran-Iraq War, there were no mass defections. None should have been expected in 2003.

The Sunni minority, too, has cross-border affinities, in its case to Syria. Saddam Hussein found refuge there after he was wounded in an assassination attempt on the Iraqi prime minister, before he participated in the coup that led to his dictatorship. The same political apparatus that keeps the minority Alawite sect in power in Sunni Syria, the Ba'ath ("Revival") Party, facilitated Sunni rule in Iraq. Ba'ath party structure is highly centralized and authoritarian, and discipline is rigid (Saddam was regularly reelected with 100 percent of the "vote"). Factional differences between the Syrian and Iraqi branches of the party precluded what leaders on both sides of the border wanted: eventual political union. In any case, the Euphrates River enters Iraq from Syria, and a corridor of Sunni habitat connects the two neighbors. That corridor, and the desert tracks beyond, have proven difficult for American forces to police. Many Iraqis and perhaps some weaponry made their way out of Iraq when it was under attack, and more recently foreign fighters have entered there on their way to Falluja and other embattled towns in the Sunni Triangle.

The military decision to end Falluja's role as a haven for insurgents in late 2004, prior to the January 30 election, had significant implications for the rest of Iraq. A massive military operation destroyed much of the city, but Zarqawi and most of his associates escaped before the fighting began. When an upsurge of terrorist activity in Mosul and Baghdad followed this campaign, it was clear that the insurgents had been repositioned rather than defeated. The election itself, however, went remarkably well, notably in the Shi'ite and Kurdish areas of the country, where turnout was high. In the Sunni areas, intimidation by the insurgents kept most voters away from the polls, so that Sunni militants could claim that the election was not legitimate. The election confirmed, however, that even in the Sunni areas the insurgents and their allies constituted a minority within a minority—a minority that had for many years controlled the majority, Sunni and non-Sunni, with violence akin to that being perpetrated daily on the streets of Baghdad and the north.

Figure 9-3 shows the area of Iraq populated primarily by Sunnis; note that the wider zone, along the Euphrates, becomes very irregular toward the northeast. In the hilly and mountainous north live most of Iraq's 4.5 million Kurds, a fraction of a landlocked, fragmented nation of perhaps 25 million partitioned by the boundaries of Iraq, Syria, Turkey, and Iran. Kurds are not Arabs and their languages are not related to Arabic, and wherever they live, they have sought greater autonomy from their rulers, whether Iraqis, Turks, Syrians, or Iranians. In Iraq this led to periodic retribution, and it is Saddam's use of chemical weapons against Kurdish villagers that left indelible impressions on the outside world. Following the 1991 Gulf War, the Kurds were given protection against Baghdad, and in their relative security they prospered as no other ethnic group in the country. Along the cultural border shown on Figure 9-3, however, there has been friction over land and rights between Kurds and Sunnis, and in addition this border area is home to smaller cultural communities such as the Turkmen and Assyrians, with whom relations are also tenuous.

Baghdad straddles the Tigris River, a vast urban agglomeration in some ways a microcosm of Iraq, with Sunni, Shi'ite, and Kurdish neighborhoods ranging from the well-off to the poverty stricken. To the east of the meandering waterway lies the slum once known as Saddam (now Sadr) City, populated by an estimated 2.5 million Shi'ites, where people were punished savagely for even minor infractions. To the west was the base of the privileged Sunnis, with key public buildings, Saddam's numerous monumental palaces, ornate mosques, statues, and other edifices of the regime. The American invasion severely damaged the city, which was for months afterward, and remains as of this writing, without a reliable water supply, power, medical facilities, or schools. It is a priority for Americans and their allies to get the capital functioning again, but the task is overwhelming in the face of unanticipated sabotage, persistent obstruction, and growing resentment.

During the year following the American intervention, the situation in Iraq deteriorated drastically, at great cost to the civilian population and considerable loss to the military. A combination of terrorism and insurgency eroded security and put social distance between the occupiers and the occupied, worsening relations and eroding trust. A scandal involving mistreatment of prisoners by United States soldiers did further damage. Bomb attacks killed hundreds of recruits waiting to join Iraq's new police and army. In effect Iraq was a failed state, and it attracted (by late 2004 estimates) between 10,000 and 20,000 foreign jihadists from near and far, bent upon driving the infidels out. The Sunni Triangle became a synonym for failure and Falluja a beacon for terrorists including the Jordanian killer of hostages, Zarqawi. The crisis in the Sunni zone overshadowed all else in the country, including very real accomplishments in the Shi'ite and Kurdish areas.

The stated objective of the American-led invasion of Iraq was to end the cruel dictatorship of Saddam Hussein, to rid the country of his political heirs (including his two rapacious sons), to develop a system of representative government, to hold elections, and to leave behind a reconstructed, democratic, economically progressing Iraq that would stand at the heart of the Muslim world as an example of the advantages of freedom and self-determination. That this Trojan Horse of democracy might be seen as something quite different by those nearby may have occurred to the planners of the operation, but no one seems to have assessed the cost of failure or indeed its potential. That cost could include a chronically malfunctioning state, a training ground for terrorists who might turn their wrath upon some of Iraq's neighbors (for example Saudi Arabia) or their countries of origin, such as France or Britain, a source and market for stolen weapons, a greater risk of regional instability, and the need for an indefinite presence of American and other foreign forces.

Whatever the outcome in Iraq, this campaign has implications for the future. Here are several to consider:

1. The United States will always, in this world of 200 states, have to deal with leaders ranging from despicable tyrants to democratically elected presidents. The destabilizing consequences of unilateral intervention at the evil end of this continuum may do more damage to the national interest than slower, multilateral pressure, unless a clear and imminent threat to nation and international community arises.

2. Given the global dissemination of weaponry it is no longer practical to attempt sociopolitical transformation of states by force. Russia's failure in Chechnya should serve as a warning that even overwhelming military force in ministates may not achieve its objectives and can generate costly cycles of retaliation.

3. Should action be taken, no transformation process in tradition-bound societies should be put on a precision schedule. Once intervention is irreversible, the commitment is indefinite. Japan is frequently and, in other ways, inappropriately cited as a precursor to Iraq as a country where American action brought democracy and prosperity. American troops are still there, 60 years later.

4. The Dayton model for Bosnia should have been considered as an option for Iraq, given Iraq's historic regionalism and the potential for particular problems in each region (de Blij, 2003). It may be too late for Iraq, but it should be an option in any future intervention in countries where discrete areas could progress under occupation toward a new order at their own pace, with eventual reintegration a "national" goal.

5. Unilateral or multilateral intervention in one state will have consequences for, and in, neighboring countries. The domino effect is likely to arise, with the potential to undermine regional stability. Even if it were to succeed in Iraq, the emergence of a democratic government would be seen as a threat by neighboring regimes unable or unwilling to transform themselves in its image.

6. The acquisition of nuclear weapons by rogue states should be prevented or constrained wherever possible, but the reality is that their dissemination is inevitable, even if Iraq had none in 2003. Interdiction should focus on the far more serious menace of nuclear-weapon acquisition by terrorist organizations. States face the risk of annihilation following first-use. Terrorist organizations do not.

7. Representative government comes in various forms, and so does democracy (the United States is in truth a federal republic, not a true democracy as critics of the Electoral College like to emphasize). Democracy takes generations to mature; homegrown varieties do better than imposed systems; it cannot usually be installed, as departing colonial powers learned.

The invasion of Iraq changed the political and cultural geography of terrorism in ways unanticipated in the immediate aftermath of 9/11. It shifted the focus from the mountains of Tora Bora to the plains of Mesopotamia, and from the terrorist bin Laden to the tyrant Saddam. It exposed the limitations of even a superpower in conducting simultaneous operations in separate and different theaters. It generated a counterinsurgency that attracted thousands of foreign fighters and provided them a training ground Usama bin Laden could only have dreamed of. It revealed disqualifying miscalculations on the part of planners who should have known their political and cultural geography better.

10

EUROPEAN SUPERPOWER?

For centuries, Europe lay at the heart of the human world. European empires spanned the globe and transformed societies far and near, for good or ill. European capitals were the nodes of trade networks that controlled distant resources. Millions of Europeans migrated from their homelands to the New World as well as the Old, creating new societies from America to Australia. Long before globalization became equated with Americanization, it was a process of Europeanization.

In agriculture, industry, and politics, Europe generated revolutions—and then exported those revolutions throughout the world, consolidating the European advantage. Yet during the twentieth century, Europe twice plunged the world into war. In the aftermath of the Second World War (1939–1945), Europe's weakened powers lost the colonial possessions that for so long had provided wealth and influence. European countries were threatened by communist parties and movements, and an ideological Iron Curtain from the Baltic to the Adriatic split the continent apart. East of it, Soviet communism dominated from its headquarters in Moscow. To the west, liberal democracy and market capitalism prevailed, but with crucial help from Washington and not before a few dreadful dictators left the scene (Hitchcock, 2002).

Western Europe's economic recovery and Eastern Europe's eventual rejection of communism were this realm's dominant events over the last half century, but another story continues to make the headlines. Europe's countries are engaged in an unprecedented experiment in supranationalism, a process of international unification and coordination in numerous spheres ranging from the economic to the political. It is an experiment that some leaders hope will ultimately lead to the formation of a United States of Europe, a federal superpower that will constitute a counterweight to the global dominance of the United States of America.

Ordinary Europeans, perhaps a majority of them, do not always share this enthusiasm for unification. But the process has gone further than could have

been imagined 50 years ago. Europeans can drive from Lisbon to Vienna without showing their passports at national borders. They can use the same currency, the euro, in a dozen countries that had their own money just a few years ago. A European Commission meets in Brussels to administer the multinational system and a European parliament sits in Strasbourg to draft European laws. And the European Union (EU) continues to expand. In 2004, it grew from 15 countries to 25, reaching far into the Eastern Europe that, until as recently as 1990, lay isolated by the Iron Curtain and encumbered by the inefficiencies of Soviet communist rule.

ASSETS AND LIABILITIES

Will Europe achieve its supranational and superpower ambitions? Look at the map, and Europe appears to be a mere protrusion of peninsulas extending westward from the great landmass of Eurasia. Read the scale, and you see how small Europe is territorially, not much larger than the United States west of the Mississippi (Texas is almost twice as large as Germany; the United Kingdom is about the size of Oregon). And Europe's territory is anything but compact. Peninsulas and islands large (by European standards) and small are flanked by seas wide and narrow. Europeans gazed at each other across straits and channels, bays and rivers from the day they settled here.

That settlement came in waves from the east as intermittent warming during the Wisconsinan Glaciation enabled the first modern humans to survive the rigors of Europe's variable climates. This invasion started slowly, but *Homo sapiens* could do what their Neanderthal predecessors could not: mix their hunter-gatherer existence with the pursuit of migrating reindeer, and adapt when the climate turned colder. Modern humans have been in Europe for more than 40,000 years, moving into Iberia and Italy during the coldest times and spreading northward when it got warmer. But the real population expansion came during the current interglacial warm period, starting about 12,000 years ago and interrupted only by the brief Younger Dryas cold spell. Wave after wave of immigrants moved from Eurasia's interiors toward the warmth of the west and the isolated security of peninsulas and islands along its coasts. They brought with them their diverse cultures and varied languages, of which the earliest are remnants now along Britain's and Ireland's western shores (Fig. 10-1). To this day some of the routes taken by these early arrivals are uncertain; no one knows, for example, how the Basques reached their present homeland on the Spanish-French border because their language has no links to any other. The Hungarians, Estonians, and Finns also arrived from yet-unknown sources and planted their languages in Europe.

LANGUAGES OF EUROPE

| 0 | 200 | 400 | 600 Kilometers |
| 0 | 100 | 200 | 300 Miles |

MAJOR INDO-EUROPEAN BRANCHES

GERMANIC GROUP

WESTERN GERMANIC	NORTHERN GERMANIC	
1 Dutch	5 Danish	8 Icelandic
2 German	6 Swedish	9 Faeroese
3 Frysian	7 Norwegian	
4 English		

ROMANCE GROUP

10 Portuguese	16 Rhaeto-Romansch
11 Spanish	17 Romanian
12 Catalan	18 Corsican-Italian
13 Provençal	19 Sardinian-Italian
14 French	20 Walloon
15 Italian	

SLAVIC GROUP

WEST SLAVONIC	EAST SLAVONIC	SOUTH SLAVONIC
21 Polish	25 Russian	28 Slovene
22 Slovak	26 Ukrainian	29 Serbo-Croatian
23 Czech	27 Belarussian	30 Macedonian
24 Lusatian		31 Bulgarian

OTHER INDO-EUROPEAN BRANCHES

CELTIC GROUP

BRITANNIC	GAELISH
32 Breton	34 Irish Gaelic
33 Welsh	35 Scots Gaelic

BALTIC GROUP

36 Latvian   37 Lithuanian

HELLENIC

38 Greek

THRACIAN/ILLYRIAN GROUP

39 Albanian

INDO-IRANIAN GROUP

40 Romani (dispersed)

URALIC LANGUAGE FAMILY

FINNO-UGRIC GROUP

41 Finnish	44 Estonian
42 Karelian	45 Hungarian
43 Saami	46 Komi

SAMOYEDIC GROUP

47 Samoyedic

ALTAIC LANGUAGE FAMILY

TURKIC GROUP

48 Turkish

OTHER LANGUAGES

BASQUE

49 Basque

Areas with significant concentrations of other languages (usually adjacent national languages)

~ Boundary between languages

Fig. 10-1

At least their familial relationships are clearer—although a recent study suggests a link between these Uralic languages and . . . Japanese.

The dominant languages of Europe today, including English and German, Spanish and French, Polish and Ukrainian, belong to the several branches of the Indo-European language family, and from their present distribution you can infer their genesis. The speakers of the Germanic languages, also including Dutch and the Scandinavian tongues, spread westward across the North European Lowland, the vast plain that extends from the shores of the North Sea into Russia between the Alps to the south and the Baltic Sea to the north (Fig. 10-2). They invaded Britain and Ireland and displaced the older Celtic-speakers there; as Europe warmed, they settled in Scandinavia and even Iceland. The Romance languages, including modern Italian and Romanian, evolved from the Latin spoken by the Tiber Valley peoples who founded and forged the Mediterranean-encircling Roman Empire and who welded their language to the tongues of its provinces. And the Slavic languages, including Czech and Bulgarian, are the languages of the east, latecomers in the Roman-Mediterranean sphere but long established in neighboring Russia.

Superimpose Europe's language map on its map of physical landscapes, and you see one major reason why this realm remains a Tower of Babel. The gentle relief of the North European Lowland (which extends into Britain as well) facilitates movement and interaction, and here lie the three historic powers and modern economic powerhouses of the realm: the United Kingdom, France, and Germany. But even here, real life is more complicated than the map suggests: major languages are not mutually understandable and minor tongues (such as Frysian) and strong dialects inhibit comprehension and reciprocity. Where the relief is higher, notably where the Alpine System prevails, physical barriers to interaction contribute to isolation and pervasive cultural fragmentation. Before its disastrous disintegration, the former Yugoslavia, about the size of Wyoming, had 24 million inhabitants who spoke 7 major and 17 minor languages. Peoples living in adjacent valleys often had no effective contact with each other, separated as they were by mountain spurs, language, and tradition.

Seen in this context, it is amazing that European integration has proceeded as far as it has. But Europe's cultural diversity is more than matched by its natural diversity, and whereas the former is a challenge, the latter presents opportunity. From the flat coastlands of the North Sea to the grandeur of the Alps, and from the moist woodlands and moors of the Atlantic fringe to the semiarid prairies north of the Black Sea, Europe contains an almost infinite range of natural environments. The insular and peninsular west contrasts strongly with the more continental, interior east. Dry-summer Medi-

EUROPE'S PHYSICAL LANDSCAPES

0   250   500 Kilometers

0   250   500 Miles

Arctic Circle

Iceland

ATLANTIC
OCEAN

Scottish
Highlands

Great
Britain

Ireland

Cambrian
Plateau

North
Sea

Scandinavian Peninsula

U P L A N D S

Baltic Sea

NORTH EUROPEAN LOWLAND

Arctic Circle

Dnieper R.

Elbe R.

Rhine R.

Sudeten Mts.

Carpathians

Crimean
Peninsula

W E S T E R N

Plateau of
Brittany

Loire R.

CENTRAL UPLANDS

Massif
Central

Alps

Dolomites

Po R.

Tisza R.

Transylvanian Alps

Danube R.

Black Sea

S Y S T E M

Cantabrian Mts.

Pyrenees

A L P I N E

Dinaric Alps

Adriatic Sea

Balkan
Peninsula

Balkan Mts.

Rhodope Mts.

Duero/Douro R.

Ebro R.

Corsica

Apennines

Tagus R.

Iberian
Peninsula

Balearic Is.

Sardinia

Pindus Mts.

Aegean Sea

Guadalquivir R.

M e d i t e r r a n e a n   S e a

Sicily

Malta

Crete

Cyprus

Longitude West of Greenwich   0°   Longitude East of Greenwich

Fig. 10-2

terranean climate in the south yields to year-round moisture on the North European Plain and cold-winter regimes in Scandinavia. Crops change from oranges and olives to fruits and vegetables to grains and potatoes. Atlantic warmth and moisture lose their effect into the continental interior, where the crop pattern changes again. Europeans have been trading for millennia; the Romans made their Mediterranean Sea an avenue of commerce.

There is more. Europe may be small, but the range of its mineral and energy resources is large. A backbone laden with raw materials extends across middle Europe from the coalfields of England to the iron ores of Silesia, propelling the Industrial Revolution when the time came (Fig. 10-3). And, as the ancient Romans already knew, there are pockets of valuable minerals ranging from copper to gold in the highlands and mountains from Spain to Scandinavia. For centuries, individual places on the European map were known for their specialized products, often based on such locally available resources. We still carry our idiomatic coal to Newcastle and advertise Italian marble in luxury homes; Sheffield was long known for steel the way Detroit was for automobiles. From resources below and atop the ground Europeans made products that were peerless on world markets: Swiss watches, Dutch cheese, Irish linen, French wines, Swedish furniture, Finnish electronics. Europe's domestic heavy industries produced trains and ships, cars and planes, trucks and tanks.

The European stage on which all this happened may be small, but it is very crowded—even after the departure of millions of emigrants headed for the New World. Americans tend to be surprised when they discover that Europe's total population is about double that of the United States, nearly 600 million in 2005. This has obvious and serious economic implications should Europe's political integration continue and its economy prosper. So far, the European economy has struggled in large measure because of the poor performance of its key component, Germany, which never fully recovered from the cost of the reattachment of former communist-ruled East Germany and also suffers from ineffective policies. As a whole, Europe's economy will also go through a difficult period following the 2004 expansion of the European Union, again involving ex-communist countries. But in the long term, Europe has the potential, at least in the economic sphere, to achieve the superpower status its leaders seek.

Crowded Europe also is one of the world's most highly urbanized realms, especially in several countries of the west where 90 percent or more of the population lives in cities and towns. Europe's great cities, from London to Rome and Paris to Athens, carry the imprints of Europe's turbulent past and tumultuous present in their historic centers, clustered, space-

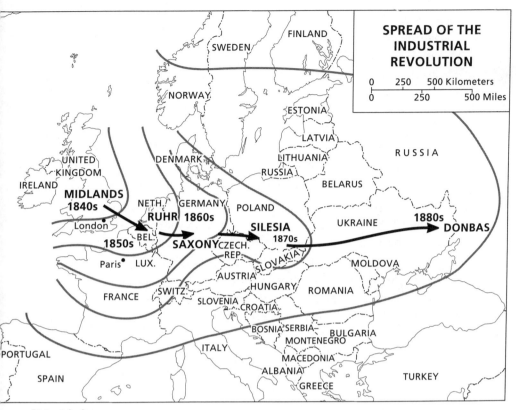

Fig. 10-3

conserving neighborhoods, and immigrant-laden outskirts. They do not display the profligate suburbanization of American cities, so that sprawl is comparatively limited and public transport more effective. But internal circulation is hampered by old, inefficient street patterns and narrow roadways. As a result, highways linking major urban centers are the scenes of some of the world's longest traffic jams. Where the modern highway meets the clogged arteries of the historic city, the slow dispersal of arriving vehicles can back up traffic for dozens of miles.

Europe's population is not only highly urbanized; it is also old. The populations of nearly half of Europe's countries are shrinking, and those of most others are growing very slowly. People living in crowded cities tend to have fewer children than their rural counterparts, but other factors also contribute to Europe's population stagnation. Later marriage, high unemployment, the cost of child rearing and other financial uncertainties, and the breakdown of religious strictures on family planning all play their role (Italy, of all countries, has zero population growth today). The implications of this for Europe's future are troubling: younger taxpayers must pay for the social services required by older citizens, and the shriveling of that taxpayer base confronts European governments with some difficult options. Combine this with the continuing influx of immigrants from all over the world, and you can see why Europe's prospects are mixed.

WHERE AND WHAT IS EUROPE?

Silly question, it might seem. Europe, of course, is the British Isles and Scandinavia and Greece and Poland . . . but maybe not Moldova or Belarus?

Welcome to a long and probably endless geographic debate. The location of the eastern boundary of Europe has been the subject of argument for many years. The key issue has always been whether Russia is part of Europe or not. When the Soviets made satellites out of much of Eastern Europe, Russia was seen by many Europeans as an external colonizer, especially when communist and noncommunist Europe went vastly different ways politically as well as economically. When the Soviet empire collapsed and Russia emerged as a fledgling democracy, there were visions of a Europe from Madrid to Moscow and beyond.

But that's just the problem. If Russia is a European state, does Europe therefore extend from the Atlantic to the Pacific, from London to Vladivostok? Not according to the cartographers who worked on the eighth edition of the *National Geographic Atlas of the World* (National Geographic Society, 2005). On page 71 their map of Asia shows a green line (a "commonly accepted" boundary, a note states) running along the Ural Mountains, then cutting across western Kazakhstan, across the Caspian Sea to turn westward along the Caucasus Mountains, and then across the Black Sea to the Bosporus. Europe's eastern border, according to this convoluted construction, largely coincides with the Ural Mountains, and there is a European Russia and a non-European Russia. It must seem rather strange for the people of the city of Ufa, just west of the Urals, to know that they live in Europe—while their compatriots in Chelyabinsk, just down the road, live in Asia.

The National Geographic *Atlas* has a lot of geographic clout, and millions of readers must have taken this notion seriously. But this version of Europe's dimensions makes no sense. If Russia is part of Europe, then Europe extends all the way to the Pacific, from Scotland to Sakhalin. If it is not, then Europe's eastern border coincides with Russia's western one.

There are many reasons to adopt this last solution. Make a list of Russia's geographic properties, and the differences between it and its European neighbors leap from the page. Russia is 100 times as large territorially as the average European country. Russia's population is nearly twice as large as Europe's largest. Russia's mainly energy-export economy is unlike that of Europe (you are unlikely to find Russian specialty products among your purchases). Russian democracy is still rickety, and two neighbors in which Moscow has residual influence, Ukraine and Belarus, retain authoritarian habits, although Ukraine is showing signs of reorientation. Russia is not even being mentioned as a potential member of the European Union. Certainly Russia's cities and cultural landscapes have European overtones, but that is not enough to conceive of Russia as a functional part of a European geographic realm.

Since we're talking about regional boundaries, what about Europe's southern border? It seems logical to regard the Mediterranean Sea as Europe's southern limit, but that's only because of what happened after Roman times. Some of the westward migrants who arrived in this part of the world took a southerly route around the Mediterranean and ended up in what is today North Africa. There they were overpowered (or ousted) by the Romans, whose provincial administrations transformed the area and integrated it into the European orbit. Rome's collapse might have merely delayed North Africa's Europeanization except for one crucial event: the arrival of Islam and the reorientation of North Africa to Mecca. By the time the French, Spanish, and Italians arrived to colonize North Africa, the Mediterranean—once Rome's Mare Internum—represented an unbridgeable cultural chasm.

Does all this matter? Absolutely. To be part of Europe means that a country can hope for access to Europe's many international economic and financial organizations, for representation in Brussels and Strasbourg, for mutual security, and for many other advantages resulting from cooperation among neighbors. The name Europe today stands for far more than a continent or a geographic realm. It also represents international opportunity and progress. Few countries would forgo the chance to join (the Norwegians and the Swiss are in a tiny minority), and some with only the remotest chance of eventually being considered for admission, such as Georgia and a potentially independent Kosovo, are among aspiring hopefuls.

Indeed, Europe has been redefined. It's no longer just a geographic locale. It's a functioning complex of countries (Fig. 10-4). How it got there is the story of this chapter.

FRACTIOUS EUROPE

In July 2003 I was asked to give a talk about Europe in England before an international audience of several hundred consisting of about an equal number of Europeans and Americans. I spoke about Germany's economic

REGIONS OF EUROPE

— European Core Boundary

1 Western Europe

2 British Isles

3 Northern (Nordic) Europe

4 Mediterranean Europe

5 Eastern Europe

| 0 | 250 | 500 Kilometers |
| 0 | 100 | 200 | 300 Miles |

Fig. 10-4

problems, France's quarrel with the United States over Iraq, prospects for the euro and EU enlargement, and the issue of a European Constitution, then very much in the news while it was being prepared, a momentous event in the EU's history. I went on too long and left no time for a Q&A session, but asked anyone with comments to come up to the lectern afterward. Soon a group of about a dozen listeners converged on me, and I could see that some of them were quite angry. "You were unfair to Germany's government!" shouted a man in the middle of the pack. Before I could answer, someone started a bitter complaint about my view of the French. "No," said the vociferous German, "he was quite right about you French. You want to run the European Union, but the British won't let you do it." In a few moments the Europeans among the group were in a shouting match with each other, no longer interested in arguing with me. When I left the room the dispute continued undiminished.

When I look at the fractured political map of Europe these days that episode comes to mind. If you don't count Europe's microstates (such as Monaco, Andorra, San Marino) or nonstate territories like Gibraltar and Kosovo, there still are 40 countries in Europe, a jumble of states creating as complex a political mosaic as any in the world. In some parts of the realm, for example along the Dutch-Belgian border, small parcels of land belonging to one country are completely surrounded by territory of the other. Europe's political map is a legacy of centuries of conflict and adjustment.

And the conflicts are not over. Basques in Spain kill members of government, judges, and policemen to stake their claim for independence. Several hundred thousand people died during the 1990s as Christians and Muslims, Serbs and Croats destroyed much of Yugoslavia's historic heritage. Until just a few years ago Catholics and Protestants in Northern Ireland fought and died as though the sixteenth century never ended. Corsicans use terror to promote their cause against the French. Other movements, many of them regional in nature, have the potential to engender strife. In February 2005, Montenegro launched a bid to negotiate secession from Serbia. Tensions over political status, cultural issues, or historic injustices (often in some combination) afflict many European countries (Fig. 10-5).

Europe has long been a crucible of culture, but it is also a cauldron of conflict. Twice in the twentieth century, Europe plunged the world into war, leading the use of weapons of mass destruction both times (gas in the first, atomic bombs in the second). Twice the combatants came out of the conflict saying "never again." But when Yugoslavia disintegrated in the 1990s, and Europeans had the opportunity to prove that they meant it, they failed. In Europe, never say never.

Fig. 10-5

REMAKING EUROPE

Against this background, Europe's success in overcoming its historic barriers and enmities is encouraging—not just for Europeans, but for states and nations all over the world. For nearly 60 years now, Europeans have been working to lower obstacles to cooperation and to facilitate the free flow of people, products, and money. This has become the world's greatest experiment in supranationalism, in which states voluntarily yield some of their sovereignty in the interest of the common good, and lessons learned in Europe have been eagerly applied elsewhere. Today there are about 40 sometimes-overlapping supranational organizations among the approximately 200 countries of the world, some of them quite consequential, such as NAFTA; others, such as Russia's almost-forgotten CIS and Africa's "African Union," of little impact as yet. For Americans, the lesson of NAFTA is that supranational cooperation has its costs as well as profits. Remember when independent presidential candidate Ross Perot spoke of that "giant sucking sound" made by American jobs escaping to Mexico? NAFTA-provoked job losses became an issue again during the 2004 Democratic primary and subsequent presidential campaigns. And NAFTA is no EU. Europeans have given up a lot more than job security to make their union work.

Lest we forget (some Europeans don't always remember), it was American generosity that got the European unification movement off the ground. President Harry Truman's secretary of state, George C. Marshall, in a speech at Harvard University on June 5, 1947, proposed a huge United States investment in Europe's recovery. It was not all altruism, to be sure: the American administration feared that communist parties would gain control over European countries west of the Iron Curtain, and helping Europe recover economically would pay huge dividends politically as well. Under the terms of the Marshall Plan (which might well be called the Truman Plan), European countries—even those behind the Iron Curtain—were asked to propose a massive self-help program that would be funded by the United States. Sixteen European countries, later joined by defeated (West) Germany, presented a plan under the title of Organization for European Economic Cooperation (OEEC). The Soviets forbade any of their Eastern European satellites from participating, so when the United States Congress funded the OEEC to the tune of $12 billion (about $130 billion in present dollars), all of it went to the West.

European enthusiasm for the Marshall Plan grew from the realization that it would have political as well as economic consequences. By enmeshing all major European states in this multinational scheme, the risk of a third war would be minimized, and Europe could set about its recovery under the security of the military North Atlantic Treaty Organization (NATO). The Marshall Plan commenced in 1948 and the protective shield of NATO took effect in 1949. Now it was up to the Europeans to convert their good fortune into lasting cooperation and to accept a rehabilitating (West) Germany as part of the mission.

The Marshall Plan lasted for four years (1948–1952) and set the stage for developments probably unforeseen even by the most optimistic European leaders (Rifkin, 2004). The Council of Europe, just a deliberative body but viewed as the forerunner to a European Parliament, was created in 1949. A crucial economic step was the formation in 1951 of the European Coal and Steel Community, removing trade barriers and establishing a common market for coal and steel among the six member states. When the Marshall Plan phased out in 1952, the momentum toward European integration never slowed. By 1957 six of the seventeen aid recipients were ready to ratify the Treaty of Rome that, in the following year, launched the European Economic Community (EEC), the so-called Common Market or "Inner Six."

Why only six? It was the old story of European divisiveness. While France, Italy, (West) Germany, and the three Benelux countries were ready to take the next step, the British felt that their future was more closely linked to the Commonwealth and they did not want to risk those ties by joining the EEC. On the other hand the British did want to keep a stake in Europe, and so they founded an organization to parallel the EEC consisting of the United Kingdom, the three Scandinavian countries, the two mountain states (Austria and Switzerland, the latter an unlikely joiner), and Portugal. This group was known as the European Free Trade Association or "Outer Seven," but it was no match for the powerful six of the Common Market. The leadership of the EEC was, to put it mildly, not pleased with this initiative. When the British changed their minds and applied for membership in the EEC, France vetoed their application. It was, so it appeared, European business as usual.

While all this wrangling was going on, certain European leaders wanted to remove the "economic" qualification from the EEC name. Europe should aspire to be more than an economic community of states; there would be other arenas of integration. And so, in 1967, the organization got its second name: European Community (EC). In 1968 the six EC members eliminated all internal customs duties and erected common external tariffs. This got the attention of the old EFTA group, and most of them, led by Britain, again applied for admission. This time the French held their fire, and in 1973 the United Kingdom, Ireland, and Denmark joined, while the Norwegian people rejected membership by referendum. The Inner Six had become "The Nine."

Behind the scenes, the fabric of the European Community got ever more intricate. The old OEEC was extended, and in effect superseded, by a broader, more international organization that now includes not only European countries but also the United States, Canada, Japan, Australia, and New Zealand: the Organization for Economic Cooperation and Development (OECD). Agreements were drawn up among EC member states ranging from agricultural policy to human rights and from monetary policy to labor regulations. One of the most important of these agreements had to do with subsidies: the richer members were obliged to help the poorer ones by contributing to a fund established to reduce inequalities within the EC. On the political front, the year 1979 saw the first elections to the European Parliament that, in that year, had 410 members charged with legislative and consultative tasks.

And the EC continued to grow. Greece was admitted in 1981, Spain and Portugal in 1986 and, after the organization was renamed once more, this time to the current European Union, Austria, Sweden, and Finland entered in 1995. Momentous meetings in prominent and not-so-prominent locales achieved ever-closer integration on numerous fronts: the Charter of Paris (1990), when East Germany was formally admitted as part of newly reunited Germany, the Maastricht Agreement (1991), charting the overall future of the EU including monetary union, the Treaty of Nice (2000), opening the door to the climactic incorporation of ten additional members in 2004 (Fig. 10-5). Now with 25 members, and with Romania and Bulgaria in negotiations to join in 2007, the European Union truly has become the New Europe.

TOO FAR TOO FAST?

You don't have to be a specialist in international affairs to realize that the expansion of a supranational organization across countries with ever-greater disparities in wealth and income, not to mention dimension and demographics, requires relaxation of the rules of membership. The old Common Market consisted of six fairly similar societies (well, five of them at least, and half of the sixth), and even the admission of the next three members kept the range of indices pretty narrow. But when Greece, Portugal, and Spain joined, the picture changed quite drastically. Not all the senior members of the EU were pleased, but soon funds were flowing from Brussels to Athens, Lisbon, and Madrid under the terms of the EU's program to help poorer countries and regions improve their infrastructures. Portugal built bridges and highways. Spain constructed a high-speed rail link between Madrid and Seville in poverty-stricken Andalusia. Greece erected a new airport near its capital.

The issue of depth versus breadth nevertheless roiled EU discussions. The longstanding objectives of the organization transcend economic integration and include common policies in foreign affairs and defense, and these seemed imperiled by enlargement. In practical terms, how could decision-making procedures originally devised for a supranational community of just six states, and modified only slightly since, function adequately for a union of 25 or more? How could the economies of poor countries such as the Baltic states be expected to meet the responsibilities and costs of EU membership after joining a fraternity organized for the wealthy? How many decades would it take for the new members to catch up with the old ones, given current economic growth rates? How much more difficult would it be for Europe to speak and act as one in the international arena when so many more voices—and votes—must be heard and heeded?

Other concerns focused on the prospect that huge numbers—perhaps millions—of workers would cross the old EU borders from the new member states and create labor troubles in the process, and again the issue split the old EU. Germany and Austria, adjoined by new members, led moves to restrict such immigration, but the United Kingdom and Ireland, more remote, opened their labor markets quickly. There was rather more unanimity on the problem of migrants seeking state benefits in the richer countries: all of the older members wanted to control this. Of particular concern were the eight million Roma (Gypsies) living in extreme poverty and facing discrimination in Slovakia, Hungary, and the Czech Republic, most of whom are unemployable and some of whom have already arrived as asylum-seekers in the West. While all the recipient countries wanted to find ways to restrict this movement, it was also agreed that the best way to do so would be to alleviate poverty in the source areas, through redirected EU subsidies. The Roma problem will arise again when Romania, home to a substantial Roma minority, joins the EU later in this decade.

A key issue confronting the EU following its 2004 enlargement centered on the way expansion affected states' relative power. The ten countries joining in 2004 had some 75 million inhabitants, of which Poland had nearly 40 million and Malta only 0.4 million—a smaller population even than Luxembourg. The issue was how to assign votes in the European Council when the largest state has more than 200 times as many people as the smallest? The older, larger members were reluctant to yield some of their power and influence to the smaller, newer upstarts, but a method had to be found to satisfy all. An early plan, hammered out in Nice, gave the two midsize countries, Spain and Poland, almost as many votes in the European Council (the EU's key governing body) as Germany, more than twice as populous. But during the drafting of the European Constitution in 2003, that plan was dropped in favor of one called the "double majority," which gave more votes to the more populous countries but required any legislation to be approved by (1) a majority of votes and (2) a majority of member countries, 13 or more until 2007 and over 14 after the pending admission of Romania and Bulgaria. When this proposal met further objection, a new plan defined a majority as consisting of at least 55 percent of member states representing 65 percent or more of the EU's population. This model would accommodate any expansion (or shrinkage) of the Union, and it satisfied all involved.

The same cannot be said for the European Constitution as a whole. Late in 2001, the then-15 EU leaders proclaimed a European Constitution desirable because, as the momentous 2004 expansion approached, the EU stood at a crossroads, a defining moment, in its history. One dimension of this defining moment was the introduction of the EU's own currency, the euro, to replace the venerable *valutas* of Europe's individual countries. Perhaps some readers will recall the amusing 1960s movie about some American tourists traveling by bus across Europe called *If It's Tuesday, This Must Be Belgium*. Well, on Monday they were spending guilders in the Netherlands, on Wednesday they were paying with francs in France, and on Thursday it was lira (by the thousands for a small lunch) in Italy. The European Monetary Union (EMU) was designed not only to facilitate the flow of money across the EU, but also to end this fiscal fragmentation. Many observers wondered if the member states' commitment to unity would extend to the abandonment of their familiar bills and coins, but the great euro experiment succeeded—to a point. Among the 15 EU members at the time the euro was introduced, 12 adopted it including all the major countries except the United Kingdom, which retained its pound sterling, and Denmark and Sweden, where voters rejected the initiative (Fig. 10-5). Now you can take your Euros from France to Spain to Italy to Germany, even to Finland and Greece. And after a slow start, the euro gained ground against the United States dollar so that, after being worth less than 90 cents following its introduction in 2002, it exceeded $1.30 by the beginning of 2005, and was poised to go even higher.

Shortly after the adoption of the euro, when the currency sank against the dollar, Europeans fretted about its weakness. But when it rose dramatically in 2004, they worried even more. This caused the prices of European products to rise on the American market, lowering sales; back home it hurt the tourist industry, slowed economic growth, and drove up unemployment. Europeans resented America's readiness to let the dollar drop, but the Europeans themselves failed to take appropriate action. The European Central Bank failed to intervene, and European leaders failed to reform their stagnant economies.

But how soon, and under what terms and conditions, should the ten new member states be offered the opportunity to replace their currencies with the euro? This was one of those "crossroads" issues that impelled EU leaders to begin constructing a European Constitution that would codify not only social but also financial responsibilities of membership. It is another one of those "breadth or depth" issues: the mostly poor new members cannot be expected to be able to adhere to the stringent fiscal EMU rules, but the core members of the Union do not want to relax these regulations (despite occasionally violating some of these themselves, as Germany did in 2003 when it exceeded EU debt limits). So the euro is likely to remain an entitlement of the "original" 15, and a distant objective for the new members.

A FEDERAL EUROPE?

The drafting of the European Constitution highlighted the divisions of opinion among EU leaders. A majority, at least among the pre-2004 membership, favors the strengthening of the powers of the European Commission and the European Parliament, a weakening of the powers of individual member states to veto EU decisions, the primacy of "European" laws over national ones, especially in the criminal justice and immigration arenas, a legally enforceable charter that guarantees a broad range of human rights, and a coordinated EU foreign policy that would serve to counter United States policies where EU and non-EU (read American) interests diverge.

The Germans and French have long been committed supporters of moves toward a federal "United States of Europe," but British leaders have been less enthusiastic. Yet although the "founding fathers" favor ever-stronger unity, they also seem reluctant to yield much of the power inherent in their countries' long-term dominance in EU decision making. French and German politicians have carried the ball: Giscard d'Estaing wrote the Constitution's lengthy and turgid preamble. But when it came to the voting system that would give all EU countries a measure of power in the EU, the heavyweights were careful to protect their turf. This gave the minority "Euroskeptics" some ammunition in their efforts to protect the powers of national governments. Self-interest, they argued, was alive and well even among avowed federalists.

To American observers, the EU's efforts to achieve consensus on something as far-reaching as a Constitution held much interest. The United States Constitution holds a special place in American life, and to witness historic allies struggle to achieve a consensus of this kind in modern times was fascinating. Quite apart from its practical side—the bureaucratic arrangements necessary to allow the EU to function through its expansion—the EU Constitution also, and more controversially, addresses sensitive issues ranging from the character of "European civilization" to Europe's "religious heritage." The draft text described the former as having "gradually developed the values underlying humanism: equality of persons, freedom, respect for reason," a construction unlikely to provoke much debate. But how to refer to Europe's religious heritage prompted much acrimonious discussion. In an earlier draft, the writers deleted specific references to Christianity as well as Greek and Roman civilization and even the Enlightenment in a compromise that sought to eliminate allusions to both religious and secular foundations of European civilization. Later, the word "religious" was inserted in a sentence that describes the "cultural, religious, and humanist inheritance" of Europe. That was not enough for representatives from Italy, Ireland, and Spain among older members and Poland, Lithuania, and the Czech Republic among the new ones, all of whom wanted some reference to Europe's Christian (or Judeo-Christian) heritage in the preamble. In truth, however, Europe, especially older Europe in EU terms, is increasingly secular; church attendance is declining and you see frequent references to "post-Christian Europe" in writings about the realm. Some framers of the European Constitution argued that the vigor of Islam in present-day Europe and Islam's historic invasions of Europe in Iberia and the Balkans justified the inclusion of "Judeo-Christian-Islamic heritage" as a defining feature of European culture, if there was going to be any reference to religion at all.

The European Constitution will have to be approved democratically by all 25 members to take effect, a process that will take time and persuasion. The British, among others, have been reluctant to put the Constitution to this test, because failure would mean a severe setback for the entire Union initiative. In late 2004 European newspapers were arguing that a "no" vote in any of the member countries might mark the beginning of the end of the whole supranational project, the first irreparable crack in an edifice that reached its apogee with the expansion just achieved. Far from moving farther toward the federal ideal, many editorials suggested, Europe was taking a huge risk in its effort to achieve consensus over its controversial Constitution (*Economist*, 2004a).

EUROPEAN GOVERNANCE

The European Union's system of governance has evolved into a complex, overlapping set of structures sometimes referred to as the "four pillars" (and occasionally called something rather less constructive). The European Commission (EC), Council of Ministers, the European Council, and the European Parliament (EP), all play roles that have changed as the EU has expanded and matured. The European Commission is the EU's executive body, long consisting of 20 members appointed by member governments for five-year terms. Its chair has the title of president; responsibilities include the formulation of legislation to be considered for approval by the second ruling body, the Council of Ministers. This council is made up of representatives from every member country and it too has a presidency—but one that rotates every six months. Many members feel that such a system of rotation may have made sense when there were six members, but not when there are 25 or more. The smaller and newer members do not want to change it (the proposed European Constitution includes a clause stipulating a 30-month presidency, however). Besides acting on legislation proposed by the EC, the Council of Ministers approves the EU budget and thus combines features of an executive as well as legislative body. Then there is the third body, the European Council, not to be confused with the Council of Ministers. The European Council consists of the heads of state of all the members, and meets twice a year. This body is needed because such items as monetary policy (including monetary union) can only be approved by the elected leaders of the states affected.

Then there is the fourth governmental organization, the European Parliament, with limited law-giving power but a huge membership set to grow even more upon further EU expansion: 732 parliamentarians in 2005 representing parties that range from Europhilic to Euroskeptic. Yes, there are elected members of the European Parliament who would, given the chance, scuttle the whole European Union project.

EU countries have "blocs" of seats in the EP reflecting their comparative populations (Germany has the largest, with 99 members). When voters in individual member states elect representatives, they can send a message to their domestic governments by voting against the incumbent party, as happened notably during the June 2004 elections. Nationalists, "greens," communists, and others with strong viewpoints can warn their domestic leaders that the next domestic election could be difficult. In 2004, for example, an anti-Union outfit called the UK Independence Party took 12 seats and 16 percent of the British vote, a warning for No. 10 Downing Street.

But the EP does not as yet have a great deal of power. Its sessions can be rowdy and contentious, but the laws it promulgates do not become EU legislation automatically. In a few areas, such as business, the environment, and budgetary matters, the European Commission has agreed to adopt EP-negotiated laws, but in general, domestic laws in the member states continue to prevail over EP legislation. In this sense, the European Parliament remains more symbol than substance.

On the other hand, the EP does reflect European organizational incoherence, the stuff of much sarcastic commentary in national media. For one thing, it meets in two places, Brussels and Strasbourg, the latter once proclaimed as Europe's future legislative capital. Visit the EP's sprawling, modernistic, mazelike building in Strasbourg and you will likely find it empty, or nearly so, since the parliament uses it only four days per month. The majority of its time is spent in its vast complex in Brussels—but the Strasbourg visits do have their purpose. Members pocket generous travel expenses and other perks, per diems for days not spent in Strasbourg, and other fringe benefits; the exposure of these practices has caused the occasional uproar. But the real budgetary excess has to do with the entire monthly intercity migration, a process many Europeans say should be scrapped.

It is thus obvious that the EU, the Commission, Council, and Parliament are parts of a work in progress, a set of structures in need of coordination and refinement. The miracle, given Europe's cultural and social fragmentation and Europeans' historic and persistent fractiousness, is that the supranational project has advanced as far and as deeply as it has.

A GEOGRAPHIC PARADOX

Even as European states converge and cooperate on their historic supranational journey, seeking the centripetal ground to cement their unification, other, centrifugal forces drive them apart. To assess the magnitude of the challenge still ahead, it is useful to see Europe's hierarchy of political entities in a seven-rank perspective (Table 10-1). The 25-member, still-expanding European Union sits atop this political-geographical ladder; such still-troublesome entities as Gibraltar, Kaliningrad, Northern Cyprus, and Ceuta and Melilla rank lowest. Between these extremes of success and failure lie five levels of formal and informal power and jurisdiction, all of which the EU must eventually accommodate.

The size and diversity of the states now comprising the EU are greater than ever, and any map showing the "new" Europe as extending from Ireland to Cyprus and from Estonia to Portugal conceals the range of economic, social, and political conditions now incorporated under the EU banner. Inevitably this leads to an "in-group" of leaders and an "out-group" of followers, a core of original states whose cooperation has advanced furthest and a periphery of countries not able to meet the same criteria of membership (Rachman, 2004). This core, however, shows its own cracks: the map might suggest that France, Germany, Benelux, and the United Kingdom should form all or most of the in-group, but in fact the British have been sufficiently ambivalent about EU participation that France and Germany have become the Union's driving force. The Schengen Agreement, for example, an early five-country multilateral treaty to drop border formalities and ease travel restrictions, included Germany and France but not the United Kingdom.

The power of the core group of states vis à vis the latecomers was evident in 2003 and 2004, when both Germany and France failed to adhere to the economic rules (in context of the growth and stability pact, limiting national debt to 3 percent of GDP) but avoided—in fact, simply voided—the associated penalties. This enraged not only the latecomers but also smaller, less powerful charter members such as the Dutch. There is no doubt about it: the EU is driven by the formidable insiders (Kagan, 2004).

The third tier consists of states that are not members of the EU, including

ENTITY	WHO THEY ARE	OTHERS
European Union	The 25	
Leaders (Core)	France, Germany	U.K., Spain
Followers (Periphery)	Poland, Slovenia, Hungary, Czech Republic	Baltics, Malta, Cyprus
Outsiders	Norway, Switzerland, Iceland	Ukraine, Serbia
Regions	Baden-Württemberg, Lombardy, Rhône-Alpes	Andalusia, Bretagne, Saxony, Tuscany
Devolutionary Pressures	Catalunia, Basque Country, Corsica, Kosovo	Scotland, Flanders, Montenegro
Fragments of History	Gibraltar, Kaliningrad	Andorra, San Marino, Liechtenstein, Monaco, Ceuta and Melilla

Table 10-1

several with strong links to the organization and others less connected. The former include Switzerland, Norway, and Iceland, all qualified but choosing to decline membership; the latter comprising an outer periphery extending from sclerotic Belarus and Moldova to Serbia and Bosnia. This outer periphery contains three candidates with prospects for EU admission before the decade ends: Romania, Bulgaria, and Croatia.

Now the geographic plot thickens. Even as states vie for admission to EU membership, the governments of many of those states are in the process of yielding various forms of authority, notably in the economic arena, to provinces, "regions," or other internal divisions. This process has been in progress for decades, and it has changed the map of Europe. France, for example, until after World War II was divided into nearly 100 *départements*, most dating from Napoleonic times (Fig. 10-6). Each département had representation in Paris, but political power was concentrated in the capital and France was a highly centralized unitary state. But today, France is decentralizing. The old framework has been replaced by 22 historically significant provinces, groupings of old départements now called regions (Fig. 10-6). These regions, though still represented in the Paris government, have substantial autonomy in such areas as taxation, economic policy, and development spending. The cities that anchor these regions, such as Lyon, headquarters of the region called Rhône-Alpes, benefit from policy because their administrations can control investment, not only within France but also from abroad. In short,

Fig. 10-6

some regions like Rhône-Alpes in France, Baden-Württemberg in Germany, Lombardy in Italy, and Catalunia in Spain (all decentralizing countries as well) have become self-standing economic powerhouses with huge concentrations of growth industries and multinational firms, and have become driving forces in the European economy.

However, such autonomy can lead to friction with the central government, which takes us to the next level of the European political hierarchy. Catalunia, among other regions and provinces in Europe, has a sometimes prickly relationship with the central government, in its case in Madrid (Fig. 10-7). With 6 percent of Spain's territory and 16 percent of the country's population, Catalunia produces 25 percent of all Spanish exports and 40 percent of all of its industrial exports. Yet Madrid had not always treated Catalunia well: no high-speed railway has ever been built to link its capital, Barcelona, to Madrid, and for some time around the turn of the century the Spanish government planned to divert water from the vital Rio Ebro away from the region toward the parched south—without adequate consultation. This angered locals and their representatives, and revived Catalunian nationalism and separatism, never far from the surface.

The story of Catalunia is repeated in many European subnational regions and provinces where productivity tends to translate into political power and in turn into separatist sentiments. But there are places where the situation is worse—far worse. Whereas mild forms of separatism in Catalunia, Scotland, and northern Italy could be negotiated (in the case of Scotland, the government in London offered the Scots their own parliament with tax-raising powers), there are places where violence accompanies separatist demands. Where this happens, as in the Basque country of Spain and on the island of Corsica in France, the process of devolution becomes menacing and destructive.

As we saw, a map of Europe showing the presence of secession-or-autonomy-minded peoples (peaceful or otherwise) is a bit unnerving. Dark rumblings from separatists in northern Italy even produced a prospective name for a putative country, "Padania." Belgium, whose capital serves as the EU's headquarters, is regionally and notoriously divided between Flemish and French speakers, the former fostering an angry Flanders-based nationalist movement. In the United Kingdom, demands for greater autonomy in Scotland and Wales led the London government to prevent the issue from spinning out of control by giving both regions the opportunity to vote on it in a referendum. Both Scots and Welsh favored the establishment of their own assemblies with limited but significant powers, and devolution was kept on a peaceful track. The aftermath of the collapse of Yugoslavia leaves major cultural-political mismatches in Bosnia, Kosovo, Macedonia, and Al-

Fig. 10-7

**AUTONOMOUS COMMUNITIES
OF SPAIN
REGIONAL GDP PER CAPITA**

City population

National average=100

- Under 50,000

	Over 140
50,000–250,000	120–139
250,000–1,000,000	100–119
1,000,000–5,000,000	80–99

National capitals are underlined

| | 60–79 |

bania; secessionist movement in Montenegro may yet lead to a breakaway
from Serbia. The Frysians in the Netherlands, the Saami in Norway, and the
Hungarians in Slovakia are among peoples who feel that their cultures are
threatened by dominant national governments. As Fig. 10-5 shows, both EU
and non-EU countries in Europe are affected, from Galicia to the Crimea.

This devolutionary spirit among so many of Europe's minorities, at a time
when the countries of the continent are trying to unify, seems to amount to a
paradox, a contradiction of the advantages of supranationalism. I raised that
question with my hosts during the height of the surge of Scottish nationalism in the 1980s, when the Scottish National Party pushed for autonomy and

demands for outright independence filled editorial pages and meeting rooms (Fig. 10-8). The answer was always the same, and it was directly related to the EU project. Scotland was an important component of the United Kingdom, a second-ranking entity with much cultural identity and adequate representation in London. But with the UK becoming just another "region" of the European Union, Scotland would sink to a lower status, with no representation of its own in a European Parliament, diminished local powers, no way to resist the imposition of unacceptable EU laws. Why should the Danes, with similar numbers and no greater cultural identity than the Scots, sit at the table in Brussels when the Scots were about to be demoted? So in a perverse way, European supranationalism served to activate Scottish nationalism.

The British government's skillful accommodation of Scottish and Welsh nationalism, and the reasoned way the Scots and Welsh accepted the government's terms, could serve as a model for other, less conciliatory separatist movements and the national administrations with which they struggle. The intractable problems of Northern Ireland, the death and destruction sowed by Basque nationalists in Spain and by Corsicans in France, the continuing violence in the Balkans and the recurrent threat of it in Cyprus all underscore the European paradox of still-intense localism against a backdrop of cooperative internationalism.

This leaves us with the lowest rung on the European political-geographical ladder: the territorial fragments of history. For all their supranational collaboration, Europeans can be awfully petty when it comes to what would seem to be minor irritants of long standing. I got a reminder of this when I was aboard a cruise ship in May 2004 after a Transatlantic crossing. I had presented a seminar on European geopolitics and the ship, after docking at Southampton, had the British colony of Gibraltar next on its itinerary. The Spanish government, in a snit over some argument with the British involving this 2.5-square-mile piece of rock with 30,000 people, refused to allow the ship to pass through its territorial waters on the way to the dock, and so the visit was scrapped. It reminded me of a time some years ago when, after visiting Taiwan on a ship, we had to sail in a raging storm to the Japanese-held Senkaku Islands before being allowed to enter the Chinese port of Xiamen. We shook our Western heads about this silly, foolish attitude, evidence, surely, of China's immature political behavior.

The imperial remnant of Gibraltar has been a bone of contention between Britain and Spain for a very long time. Its inhabitants have voted against independence and against incorporation with Spain; the Spanish have blocked Gibraltar's airport and shut its border gates. Contrary to popular impression, Gibraltar does not overlook the narrowest part of the strategic Strait

THE BRITISH ISLES

POPULATION

- • Under 50,000
- • 50,000–250,000
- • 250,000–1,000,000
- • 1,000,000–5,000,000
- • Over 5,000,000

— Gas/Oil pipeline
♦ Gasfield
⚓ Oilfield

— Maritime Boundaries

National capitals are underlined

0   50   100   150   200   250 Kilometers
0   25   50   75   100   125   150 Miles

STATFJORD   BRENT
ALWYN   NORWAY
Shetland   Bergen
Islands   60°
FRIGG
Stavanger
Orkney
Islands   PIPER   North
5°
Inverness   FORTIES   Sea
Outer   Aberdeen   COD
Hebrides   SCOTLAND   EKOFISK
Dundee   UNITED   DAN
Glasgow
Edinburgh KINGDOM   55°
Newcastle-
upon-Tyne
Londonderry
55°
NORTHERN   Carlisle
IRELAND   Middlesbrough
Belfast   Isle   PLACID
of   York Kingston-
Man Barrow Leeds upon-Hull
Irish Sea   Manchester   INDEFATIGABLE
Dublin   Liverpool   Sheffield   LEMAN
ENGLAND   BANK Amsterdam
IRELAND   Leicester   NETHERLANDS
Limerick   Birmingham Coventry Norwich
Northampton   Cambridge
Cork   Oxford
WALES
Cardiff Bristol
St. George's Channel   London   Dover
Southampton Brighton   Brussels
Exeter   Portsmouth   BELGIUM
Plymouth   50°

ATLANTIC
OCEAN

English Channel   FRANCE

Paris

10° Longitude West of Greenwich   5°   0°

Fig. 10-8

of the same name: it is situated off the western entrance. Prosperous and multicultural, the locals are understandably reluctant to change the status quo, and the British give them the last word. Spain's bottom line is that it wants Gibraltar returned to Madrid's jurisdiction. The standoff, punctuated by quarrels arising over minor incidents, roils relations between the two EU members and affects cooperation involving much bigger issues such as fishing regulations and security.

Spain has reason to view Gibraltar as an anachronism in modern Europe, but the Spanish have their own outposts, Ceuta right across the Strait of Gibraltar and Melilla across the Alboran Sea, along with several small islands off the Moroccan coast. Talk about Spain yielding these exclaves to Morocco, and you hear echoes of British colonial policy. The locals should decide, and they prefer to stay with Spain.

The nominal independence of the tiny entities (microstates) of Monaco, Liechtenstein, San Marino, and Andorra (which is larger than Malta, a full-fledged member of the EU, go figure) precludes similar jurisdictional problems arising from these fragments of history, but there is one territory that does have the potential to cause difficulties in the future: Kaliningrad. The map of Europe shows this exclave of Russia, about the size of Connecticut, facing the Baltic Sea between Poland and Lithuania. Centered on the fortified city founded in 1255 named Königsberg, that joined the Hanseatic League in 1340 and became the residence of Prussian dukes after 1525, Kaliningrad got its new name and regional borders after 1945 as part of the Potsdam Agreement following World War II. The then–Soviet Union's communist regime expelled the German population, gave Kaliningrad the status of oblast, and made this strategic corner one of the most heavily militarized and industrialized components of the U.S.S.R. When the Soviet Union collapsed, Russia inherited an irrelevant military complex, failing manufacturing plants, inefficient collective farms, a rusting fishing fleet, a bleak "socialist" city and a number of stagnant district capitals.

Kaliningrad, however, may not be so immaterial in the future. It still gives the Moscow government a warm-water outlet to the Atlantic; it now lies between two new members of the European Union, its population of about 1 million includes a substantial Lithuanian minority, and were it not for the Potsdam Conference, Poland would undoubtedly have been given all or most of Königsberg as part of its award of East Prussian territory at the end of the war. While neither Poland nor Lithuania is likely to lay claim to any part of Kaliningrad, the map suggests another possibility: that Russia will seek a route of egress via Belarus and either Lithuania or Poland. The northwest corner of Belarus lies only about 60 km (40 mi) from the southeast corner of Kaliningrad, and there has been talk of a corridor or transport artery along the Lithuania-Polish border. Russia's troubled relations with the autocratic regime of Belarus, however, have not been conducive to such a prospect. Nevertheless, Kaliningrad has the potential to create friction (Stanley, 2001).

Uncertainty also surrounds the future of Cyprus. In early 2005, this island in the eastern Mediterranean, much closer to Turkey than to Greece but with a Greek majority, remained divided between the legitimate (in the eyes of the international community) Republic of Cyprus, on the south side of the island, and the Turkish Republic of Northern Cyprus, with under 200,000 inhabitants and 40 percent of the territory, in the north. When Cyprus joined the European Union in May 2004, only the south side with its 900,000 residents was admitted. A last-minute effort by the United Nations and European intermediaries to reunite the two sides in a loose confederation collapsed, in large measure because of a dismal failure of leadership on the Greek side resulting in a defeat of the plan. The notion of a Turkish minority being part of a country's accession to the European Union was attractive to European leaders who saw this as a step toward the eventual admission of Turkey itself. Instead, the problem of a divided Cyprus festers, and the risks this entails are serious. Here is another of Europe's trouble spots, where violence has accompanied the evolution of the cultural landscape.

THE PERILS OF EXCLUSION

When the European Union expanded to 25 members in 2004, a new geographic boundary appeared on the European map. From the shores of the Gulf of Finland, where Estonia meets Russia, to the head of the Adriatic Sea, where Slovenia borders Croatia, extends an old-style boundary of the kind no longer seen within the EU itself, with visa requirements, passport controls, checks on cargo, and sometimes time-consuming legal paperwork. This intra-European boundary separates fortunate Europeans, ensconced in the EU, from outsiders wanting, in most cases, to get in. It is a boundary set to change: Romania and Bulgaria were in accession discussions even as the 2004 expansion took place, and Croatia was trying to start negotiations as well. But in other areas, this is a tough, divisive line. Where the EU boundary separates Belarus and Poland, it truncates the hinterlands of places like Hrodna and Brest. The boundary between Ukraine and Poland partitions a historically integrated region that encircles the Ukrainian city of Lviv. Travel from southeastern Poland into western Ukraine and you are likely to see few reasons why Poland should be "in," and Ukraine "out." Go farther east, however, and the causes are clearer. The government in Kiev and its eastern Soviet-era economic and political baggage make accession a distant prospect.

But there is another side to this. Poland's entry into the EU makes it vulnerable, from a corporate viewpoint, to wage increases not occurring in Ukraine. Already, some German firms are opening factories on the Lviv side of the border, where per-hour labor costs are one-fourth of those in Poland. Foreign investment in the Lviv area is rising faster now that EU boundary controls keep workers from crossing it, ensuring a steady labor supply. Even in Belarus's town of Brest, the Minsk government has set up a so-called economic free zone in recognition of its location on the most direct road and rail routes from Warsaw to Moscow. EU companies see not only low wages but also the open border between Belarus and the large markets of Russia, so they view Brest as a potential profit center. But the regime in Minsk makes matters difficult, to say the least, for private enterprise, putting strict limits on the number of workers a private company may employ and altering tax rules overnight.

Belarus appears destined to remain on the "wrong" side of the EU boundary for a long time to come, as does Moldova, wedged between Ukraine and Romania, and by many criteria the poorest country in Europe. But in the autumn of 2004, a crucial sequence of events changed the political landscape in Ukraine, raising the previously unthinkable prospect that this, territorially the largest of all European states, may eventually join the EU. Industrialized eastern Ukraine has a large Russian minority and is closely linked to Moscow; more rural western Ukraine is the cultural heartland of the Ukrainian nation. In an election for president, a pro-Russian candidate (supported on the stump by Russian president Putin himself) initially defeated an opponent who had strong support in the west, but the election was fraudulent and crowds took to the streets of the capital to demand a repeat under international supervision. In the rerun, the more nationalist candidate won, but the electoral map had ominous overtones: the pro-Russian candidate won in all districts in the east, and his opponent in all of the west. The prospect arose of a split of the kind that happened in the former Czechoslovakia, but the new president moved quickly to restore relations with Moscow and to reassure the eastern part of the country that its concerns would be addressed. Nevertheless, it was immediately clear that Ukraine would tilt more toward Brussels than Moscow, and early in 2005 its government issued an official expression of interest in future discussions with the European Union over "issues of mutual interest."

If Romania and Bulgaria can be admitted to the EU, as they are scheduled to be before the end of this decade, then Ukraine certainly has a case. When this happens, Greece, long the geographic EU outlier, will finally be land-connected to the organization of which it is a member. This realignment will define much more clearly the Balkan group of outsiders hoping to join the EU some day: the

Fig. 10-9

remnants of the breakup of Yugoslavia (Croatia, Bosnia, Serbia-Montenegro, Macedonia, and the in-limbo territory of Kosovo) plus Albania (Fig. 10-9).

In this cluster of Balkan countries the EU will face what may well be the most difficult of all challenges in bringing coordination and stability to its ever-widening circle of member states. Ethnic and cultural tensions run deep here; devolutionary pressures persist; a major Muslim presence creates particular problems; poverty and distrust go hand in hand. Neighboring EU countries have strong interests in the course of events: the Hungarians in

Serbia, where there are substantial Hungarian communities, the Greeks in Macedonia, historically a Greek sphere of influence. The peace of Europe has foundered before in this fateful mosaic of cultural discord, and it is by no means certain that even the EU will mitigate these enmities.

AN ISLAMIC EU MEMBER?

Long before the momentous expansion of 2004 took place, still another neighbor on the outside of the EU boundary signaled its interest in joining: Turkey. Indeed, when Turkey made its first request for accession talks, it was joined by Morocco and the EU was still called the European Community. Morocco's request was quickly denied, but Turkey's representatives were invited to begin preliminary discussions. Nearly 20 years later, discussions are still going on, but unlike Romania and Bulgaria, Turkey has been given no target date for membership.

Turkey's candidacy has, in fact, divided EU members and leaders for many years. If the EU does not include Ukraine or Serbia, should it incorporate Turkey? If criteria for membership include the humane treatment of minorities and the absence of a military role in government, how can Turkey even be considered? If one country defies the entire international community and recognizes a "republic" established by force of arms, should that country be allowed to apply for EU membership?

On the other side are those who are prepared to bend the rules for Turkey because its membership would confirm the capacity of the EU to encompass even greater cultural diversity than Europe itself contains, put an end to allegations that the EU is a "Christian Club," and create a bridge between the European and Islamic worlds. There are also hopes that Turkish involvement in the EU would help integrate Islamic (many of them Turkish-Kurdish) communities into the mainstream of European life (*Economist*, 2004b).

The discussions relating to Turkey's candidacy have already had major results in Turkey itself, ranging from enhanced freedoms for the country's Kurdish minority to reduced military involvement in government, from the abolition of the death penalty to increased legal protection for women. But, as Turkey's long-dubious Greek neighbors like to point out, the Turks have a long way to go. In the EU itself, Turkey's improving prospects are cause for some introspection as well. Turkey's population of 70 million would rank it second in the EU; its growth rate (1.5 percent in 2004) is faster than the world average at a time when Europe's overall population is declining, so that by the end of the next decade Turkey would be the EU's most populous state, having overtaken Germany. "The Islamic tail wagging the Christian dog," wrote an observer of Turkey recently, but the fact is that European Christianity itself is declining even as Islamic fervor is strengthening. To many Europeans, the entire initiative involving Turkey is further evidence that the bureaucratic elite of the EU is out of touch with the people in general, but barring some catastrophic failure of the EU itself, Turkey appears to be headed for incorporation (*Economist*, 2004c). Look at the map and marvel: "Europe" would then adjoin Iraq, Syria, and Iran.

ALLY OR ADVERSARY?

To answer our original question, Europe is a superpower already—an economic superpower. The EU's giant economy is a formidable competitor for the United States; the European market (the continent as a whole or the EU only) is much larger than that of the United States if not as rich per capita, and barring poorly governed Germany, the European economy has continued to grow at a healthy rate. Add Turkey to the mix, and the "European" population is over 650 million, more than twice as large as America's.

But Europe is no military superpower. European investment in the armed forces remains comparatively small. Europeans have been content to let the United States dominate NATO (which also expanded in 2004, though far more quietly than the EU: in March of that year, no fewer than seven countries joined NATO including the three Baltic states, Romania, and Bulgaria). Europe's dismal failure to stop the genocide in collapsing Yugoslavia in the early 1990s was followed by joint American-European intervention, chiefly in the form of high-altitude bombing, in Serbia and Kosovo in 1999, without the kind of UN Security Council authorization Europeans soon thereafter sought so fervently in the case of Iraq. Throughout the Cold War the Europeans were content to leave their security in the hands of the United States, and NATO was the bulwark of that security. After the Soviet threat dissipated, Europeans began to complain about United States hegemony and unilateralism. Europe's military impotence was starkly clear during the Serbia campaign, when American intelligence, equipment, and strategy determined the course of action and the European contribution was little more than symbolic. Surely this was to be expected: Europeans had the luxury of their political realignments and economic successes without having to worry about—or finance—a military apparatus that would give them the power to match the Americans and to influence policy.

So now the Europeans want a counterweight to American military power, but their only recourse is through international institutions and alliances among EU members to obstruct United States unilateralism and hegemony.

Germany and France sought to obstruct American policy toward Iraq by using the same United Nations they had—jointly with the United States—circumvented in Kosovo. This led a frustrated American defense secretary to complain impolitely that an "Old Europe" was impeding policies a "New Europe," such as Poland and Romania, would recognize as salutary. In 2003, United States–European relations approached a low ebb, with the United States president declining to meet the German chancellor and anti-French jokes becoming the grist of the American comedy mill.

Clearly, Europe is not on the way to becoming a military superpower to challenge or influence United States dominance, and European actions were born of frustration as much as perceived moral certitude. In the constitutional discussions, Europeans even had difficulty agreeing on the portfolio of an EU foreign secretary, an equivalent to the American secretary of state to coordinate EU foreign policy; the notion of a potent all-European armed force under some form of joint command remains a mirage. The EU includes nuclear powers (the UK and France) and a wide range of national militaries, and the NATO treaty is what binds them. The United States continues to be that organization's paramount power. Only the dissolution of NATO would change this landscape.

It is therefore evident that Europe is and will be America's ally, not a military counterweight. For all the divisive language heard (and actions taken) during the past half decade, the United States and Europe have far more in common, in shared global objectives as well as cultural foundations, than divides them. Indeed, cultural and political divisions within Europe, and within America, in some ways appear stronger than those separating us from each other. When Europeans and Americans are asked to summarize the goals they have for this world, their responses are remarkably similar. The difference lies in the ways to achieve those goals.

In this context it is worth revisiting the geographic issue of Europe's borders. Surely Americans and Europeans share the hope that Europe's great experiment will succeed, that its internal boundaries will soften further and that the EU's external border will move inexorably eastward to incorporate not only Romania and Bulgaria but also a progressive Ukraine, a democratic Belarus, a stable Georgia and, ultimately, a reformed Russia, so that the argument over Europe's geography may be settled by a simple, hopeful phrase: Europe reaches across Eurasia from Atlantic to Pacific.

11

RUSSIA:
TROUBLE ON THE EASTERN FRONT

In this fast-changing world, what a difference a decade makes. In the late 1980s the Soviet Union still was one of the world's two superpowers, a colonial empire extending from the Baltic Sea to Central Asia, a communist enforcer in control of most of Eastern Europe, a nuclear-armed behemoth capable of global destruction. Ten years later, its empire disintegrated, its ideology discounted, and its army in disarray, Russia, the imperial cornerstone, was struggling to reorganize as a democracy and to reestablish a position of consequence on the world's geopolitical stage. But by the middle of the first decade of the new century, Russia's major contest was not with other giants on that stage, but with tiny Chechnya within its own borders. What remained of its armed forces were not at war in some remote Asian frontier but inside Russia itself. Thousands of Russians had died violently, many in terrorist attacks in the capital, Moscow. The cost of this tragedy far exceeded the lives lost and property destroyed. It also compromised Russians' efforts to sustain their march toward democracy, openness, and the rule of law, and brought widespread fears of a return to the authoritarianism that had marked Russian and Soviet governance for so long. Yet a Russia with representative government, whose armed forces are under civilian control and whose laws function effectively, is key to the stability and future economic and political integration of Eurasia.

GEOGRAPHIC PROBLEMS OF A TERRITORIAL GIANT

Not only is Russia the world's largest country territorially: it has more neighbors than any other state. Geographically, nothing is simple when it comes to Russia, and so it is with this set of neighbors (Fig. 11-1). By virtue of its exclave of Kaliningrad, Russia has Poland and Lithuania as European neighbors, as well as Finland, Estonia, Latvia, Belarus, and Ukraine. That makes seven neighbors in Europe alone, and Russia has issues with almost all of them. In the case of Lithuania, Russia wants free transit for Russian freight

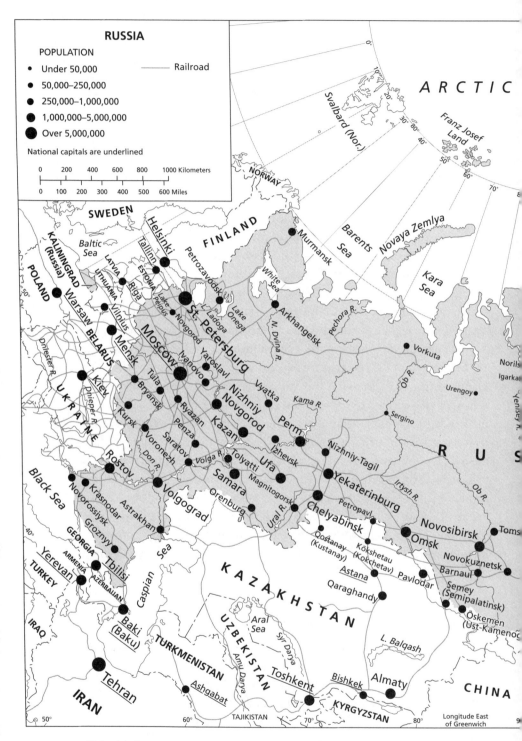

Fig. 11-1

and military traffic (and no inspections) to Kaliningrad. In Estonia and Latvia, which, like Lithuania are former parts of the Soviet empire, Russia wants official status for the Russian language, spoken by large minorities who moved here from Russia during communist times. In Belarus the situation is unusual: that country's Soviet-style, authoritarian ruler has repeatedly pressed Moscow for closer association, even formal union, between the two countries, but the Russians have so far declined the invitation. With a large Russian minority in Ukraine, Moscow has meddled dangerously in the politics of that large and culturally divided country, once the Soviet Union's second-ranking component in terms of population as well as economic output. Today, many Ukrainian leaders want to see their country join the European Union, but others see their future in closer ties with Russia.

Along its southern border between the Black and the Caspian Seas, Russia borders two former dependencies, Georgia and Azerbaijan (Fig. 11-1). But that simple statement belies a complicated geographic situation that is at the root of many of present-day Russia's most dangerous problems. As a more detailed map shows, Russia's borderland in fact consists of a tier of internal "republics" designated to recognize the non-Russian ethnic composition of their populations, and it is these republics, including Chechnya, that border Georgia and Azerbaijan (Fig. 11-2). This is the region Russians refer to as Transcaucasia, and here live ethnic minorities with memories of Russian subjugation and oppression who would have wanted the same independence given to Georgia and Azerbaijan (and their neighbor Armenia) when the Soviet Union collapsed. But that was not to be, and as a result this region became a cauldron of conflict involving not only the Muslims of Chechnya, who oppose Moscow, but also those of Ingushetiya, who have tried to avoid taking sides, the North Ossetians, who generally support Russia, the Balkars of Kabardino-Balkaria, accused by Stalin of pro-Nazi sympathies during World War II and exiled en masse, and literally dozens of other ethnic groups with turbulent histories. Meanwhile, across the border in Georgia, pro-Russian Abkhazians and South Ossetians defy the government in Tbilisi even as Russians maintain military bases on Georgia's soil. And while oil continues to flow from Azerbaijan to Russian terminals on the Black Sea, pipelines are being laid to divert much of it via Armenia to the Mediterranean coast of Turkey. Transcaucasia makes Russia's other borders look uncomplicated by comparison.

Russia's four Asian neighbors also form a contentious group: Kazakhstan, Mongolia, China, and North Korea (Fig. 11-1). When the Soviet Union collapsed and Kazakhstan became an independent state, several million Russians found themselves on the wrong side of the border in northern

Fig. 11-2

Kazakhstan, where Russia's space port and launching facilities are also located. The new Kazakh government negotiated agreements to allow these
operations to continue, but the Russified north presented far more difficult
problems because this area had, in effect, become part of the Russian sphere
(as the transport systems on the map confirm). While many Russians emi-

grated back to Russia, others agitated for secession, prompting the Kazakh government to move its capital from the Kazakh heartland in the southeast to Astana in the Russified north, underscoring its claim to the entire country. Immediately to the east lies Mongolia, a Soviet satellite during the communist period, then closely associated with Russia after 1990 (Russian was the main foreign language spoken here) but now reorienting toward China, currently its largest trading partner by far. This is changing the significance of the Russian-Mongolian boundary, once merely an administrative device but potentially a marker between Russian and Chinese spheres of influence. Still farther to the east, Russia shares a lengthy and historically contentious boundary with China that has been the scene of territorial disputes and cross-border skirmishes. These issues were settled in recent years through negotiations between the two governments, now on better terms. But another issue is emerging: massive cross-border migration by Chinese traders and workers into the Russian Far East (Davis, 2002). And in the farthest reaches of its eastern frontier, Russia has a short but consequential border with North Korea. North Koreans are escaping their tyrannical rulers by crossing into China, and from there some are reaching Russian soil. Indeed, the local regional government wants to encourage this immigration because, as we will see later, Russia's Far East is losing population. But Russia, once North Korea's ideological ally, has other worries, lying as it does within reach of North Korea's rockets, and potentially, nuclear weapons. Hence Russia is one of the six members of the team of nations seeking to temper North Korea's nuclear ambitions.

Russia's boundaries enclose a country that, for all its continental size, is very nearly landlocked. The czars of old were in a constant drive to push Russia's limits seaward. Peter the Great wanted to make Russia a maritime as well as a land power and built St. Petersburg, his new capital, on Russia's window to the Baltic Sea between Finland and Estonia. Catherine the Great sent her armies to the shores of the Black Sea and into Transcaucasia, but her real objective was an outlet on the shores of the Indian Ocean. The British and the Turks denied her that goal, so that Russia west of the Urals depended on the seasonally ice-blocked Baltic and the narrow, Turk-controlled Bosporus and Dardanelles for outlets to the sea. True, Russia's eastward expansion to the Pacific Ocean gave it a major port at Vladivostok, but this was no practical alternative to a nation concentrated thousands of miles across Siberia to the west. And Russia's borders create another problem obvious from the map: for all its bulk, the country is almost entirely confined to high, cold latitudes under Arctic influences much of the year (see Figure 4-4). Grain shortages during the Soviet era drove communist planners to expand farm production through irrigated megaprojects in the republics, but even then Moscow had to depend on costly imports from the west. Now the Soviet Union's breadbasket, Ukraine, is an independent country and Russia's climatic quandary is even more pronounced. As a Russian geographer once said to me, "our borders have never been our friends."

Crossing those borders overland during the Soviet era was a daunting experience. I rode a bus from Helsinki to (then) Leningrad in June 1964, my first field trip to the U.S.S.R., and I learned that the boundary on the map was in fact a wide zone on the ground. We reached the Finnish border just before dark, and formalities were quick and courteous. We reboarded the bus and proceeded into a treeless corridor, to be met by a carload of uniformed, armed guards who escorted us several miles down the road. In the distance was a patch of bright light, so bright that when we reached it, it was as if the bus had been driven into a surgical theater. There we were ordered off the bus and all luggage and cargo was unloaded. Passengers were separated into three groups, Soviet, European (I was traveling on a Dutch passport), and others, including a group of Canadian and American academics. Every piece of luggage was examined in minute detail, and we were physically searched in ways that make the current airport-security procedure seem casual by comparison. Then we were instructed to sign documents stipulating that we were not carrying items ranging from books and "documents" to weapons and "propaganda." The entire operation took about three hours, and I wondered how long the wait would be when a line formed. "Never a queue," said the English-speaking guard handling the North Americans. "Only three buses a day and maybe five cars." I thought about that as heavy armored gates swung open and our driver headed into the darkness of the road to Vyborg. The main road between the capital of a neighboring country and the main port of Russia, and the daily traffic amounted to fewer than ten vehicles. Soviet borders were barriers indeed.

A VAST REALM

Even after the loss of its 14 dependencies, Russia remains the world's giant state territorially, nearly twice as large as the next-ranking country, Canada, and with 13 neighbors (no. 2 Canada has one). From the volcano-studded Kamchatka Peninsula in the Russian Far East to the great port city of St. Petersburg in the west, the country stretches across 11 time zones. A television program called *Good Morning Russia* airing in Moscow at 7 A.M. local time would be seen in Vladivostok at dinner time. Russia's northernmost Arctic-sea islands lie north of 80 degrees latitude; its southernmost sliver of land adjoins Azerbaijan in Transcaucasia, still above 40 degrees. Putting this in North American terms, all of Russia lies north of the approximate latitude of Boston.

It is worth taking a moment to look at the map of world climates (page 88) to see just how cold Russia really is. Neither distance nor mountains protect it against invasions of Arctic air. Almost all of it is dominated by D climates, which in the west have a short warm summer that diminishes eastward, creating the frigid conditions for which Siberia is a synonym. Not until maritime influences moderate the climate along the Pacific coast do Siberian conditions let up. Looking at the map of world population distribution (page 98) we can see that most of Russia's over 140 million people cluster in the mildest corner of the country, the west, and in a ribbon along the southern margins of Siberia, where the Trans-Siberian Railroad links cities and towns and connects the populated west to the sparsely peopled Far East.

Just as Americans use geographic references such as "Midwest" and "Great Plains," so do Russians refer to their vast country's broad physiographic regions. The great divider of Russia is the Ural Mountains, the Appalachians of Russia but located farther into the interior (Fig. 11-3). The Urals extend from the Arctic Ocean, where they rise above the water as glacier-carrying Novaya Zemlya Island, to (and beyond) the desert border with Kazakhstan. West of the Urals, in the perception of many people, lies "European Russia"; to the east, therefore, lies something else, though the cultural landscapes of Russian towns to the east of the Urals are remarkably similar to those of the west. In any case, the heart of Russia, its core area in geographic lingo, lies on the Russian Plain, an extension of the North European Lowland, cooler and drier but still productive agriculturally. At the center of it is Moscow, on its short Baltic coast lies St. Petersburg, and crossing it is the great Volga River, flanked by industrial cities all along its course.

Siberia begins on the eastern slopes of the Urals and does not end until the shores of the Bering Sea, but its relief does change from west to east (Fig. 11-3). Westernmost Siberia, region (3) on the map, has comparatively low relief and is drained by a major river system, the Ob-Irtysh, whose gradient is so slight that Soviet engineers talked of reversing it to irrigate farmlands to the south (Lincoln, 1994). In this forbidding, forested, frigid countryside lay many of the prison camps of the infamous Soviet gulag, in which, historians estimate, between 30 and 60 million inmates perished during the seven decades of communist rule (Remnick, 1993). Along Siberia's more livable southern margin lie cities such as Omsk and Novosibirsk, strategically crucial during World War II when much of Soviet industry was shifted eastward, across the Urals and away from the Nazi advance.

At the eastern margins of the West Siberian Plain, the relief changes quite dramatically, especially in the south, where jumbled mountain ranges rise from the plain. In the north, Siberia takes on the character of a rugged plateau. Here the Trans-Siberian Railroad passes through narrow valleys and hugs the walls of steep gorges, eventually emerging from this rough terrain to reach the key city in the area, Irkutsk, gateway to Lake Baykal. This freshwater lake lies in a rift valley similar to those of East Africa's Great Lakes, but Lake Baykal, nearly 400 miles (640 km) long and averaging 30 miles (50 km) in width, is even deeper, reaching more than 5,300 feet (1,620 m) in depth. By some calculations Lake Baykal contains one-fifth of all the freshwater on the Earth's surface, and its unique ecology attracts an endless stream of researchers from all over the world to study it.

Now comes Russia's vast, forested, mountainous east, region (6) on the map, lower in the Yakutsk Basin and higher in the spectacular Kamchatka Peninsula, the country's most geologically active zone. Don't expect to drive to this earthquake-prone, volcanic slab of tectonic plate (ironically, northeastern Russia is geologically part of the North American Plate!), because there are no connecting roads. The people who share this peninsula with more than 20 active and over 100 dormant volcanoes live as though they were on an island, fishing for a living and boating or flying to the mainland when the need arises.

As Figure 11-3 shows, the Russian Far East incorporates one real island, named Sakhalin, and this is an important component of this region's physical as well as cultural geography. From the mid-nineteenth century on, the Russians and the Japanese repeatedly fought over Sakhalin Island, and not until the end of World War II was Soviet control confirmed. When the Russians held it, they used Sakhalin as a penal colony (the great writer Anton Chekhov in one of his books described the terrible conditions under which prisoners lived), but during Soviet times Sakhalin became an increasingly important source of fuels ranging from oil in the north to coal in the south. In post-Soviet years additional finds of oil reserves have made Sakhalin Island a key constituent of the commodity-based Russian economy.

Russia's enormous size bestows it with a large inventory of natural resources, among which oil and natural gas have been the key money makers during the post-1991 period. The Soviet Union's first dictator, V. I. Lenin, was determined to speed his country's industrialization, especially its heavy manufactures. For this the U.S.S.R. contained almost everything it needed, from coal to iron ore and from other metals to alloys. When World War II loomed, this resource base allowed the Soviets to build their own weaponry with which to defeat the German invaders, and afterward Russian factories

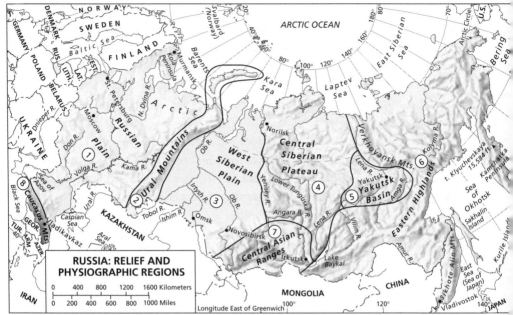

Fig. 11-3

produced most of what the country required, from automobiles to railroad cars and from tractors to passenger planes. When, during the 1950s, the Russians launched the first Earth-orbiting satellite and a Russian cosmonaut was the first in space, it was a homegrown project that astonished the world, proving, it seemed, the superiority of the Soviet system.

But for all its size and large population (third largest in the world while the U.S.S.R. lasted), economic interaction with the rest of the world was very limited. Soviet products did not appear on international markets; Russian automobiles were not seen, except as a curiosity, on foreign streets. The state enterprises of the command economy produced goods at costs and quality levels that would have made them uncompetitive in any case, so that the largest volume of exports was not consumer goods but weaponry, sold by the Soviet government to allied regimes. The economic geography of the Soviet Union resulted from assignment, not efficiency—certain cities and towns were allocated production tasks based on criteria other than cost. The potential for corruption in such a system can only be imagined. When the Soviet Union collapsed and Russia faced integration into the world economy, the massive sale of commodities—oil and natural gas—became the major source of badly needed external revenues. Russia was fortunate to possess substantial energy reserves from its share of the Caspian Basin in the west to Sakhalin Island in the east, with ready customers including Europe, Japan, and China. But overdependence on a single commodity entails risks and skews development. It opened the way to "crony capitalism" as high-level insiders bought up state enterprises, tax collections faltered, factories failed, and, in 1998, the government defaulted and the economy crashed, driving away the first wave of investors. A vast reservoir of natural resources alone is not enough to ensure this (or any other) country's prosperity.

SOVIET LEGACY, RUSSIAN CHALLENGE

In the political-geographic arena, too, the problems Russia confronts are daunting. Organizing the administration of so vast, remote, and isolated a realm has posed a historical challenge for Russian czars, communists, and democratically elected leaders alike. Expansion was one thing—Russian armies were able to penetrate deep into Central Asia, and Russian colonists claimed Alaska and built their southernmost fort near San Francisco Bay—but consolidation was quite another. When United States secretary of state William Seward offered to buy Russia's Alaskan holdings in 1867, the Russian government quickly agreed, because these outposts were becoming more trouble than they seemed to be worth. In truth, successive czarist rulers never established a satisfactory administrative structure for the numerous peoples, Russian and non-Russian, under their control. Europe's democratic revolution passed Russia by, and its economic revolution touched the czars' domain only slightly. Most Russians, and tens of millions of non-Russians under czarist domination, faced exploitation, corruption, subjugation, and starvation. When desperate rebellions erupted in 1905 and full-scale revolution broke out in 1917, there was no political framework to hold the state together (Shaw, 1999).

Thus it fell to the communist victors in that revolution, the Bolsheviks, to design the regional framework that would constitute the Soviet Union. The basic structure created 15 "Soviet Socialist Republics" (SSRs) among which the Russian Soviet Federative Socialist Republic (RSFSR) was first among theoretical equals. The other 14, from the Estonian SSR on the Baltic to the Tajik SSR bordering Afghanistan, were designed to accommodate non-Russian peoples who had fallen under Russian domination during czarist times. That system, of course, fell apart when the Soviet Union collapsed. More durable was the framework the communist planners laid out within Russia itself, because this is what the post-Soviet leaders inherited, and with which they had to work in their effort to transform the vast country from communist dictatorship to democratic consensus.

It is useful to take a look at this convoluted Soviet system, because it contains the seeds of the troubles post-Soviet Russia has faced in the years following 1991 (Shaw, 1999). The Soviets, always mindful of status and hierarchy, divided the RSFSR into internal "republics," autonomous regions (*okrugs*), provinces (*oblasts*), and territories (*krays*). The "republics," like those beyond Russia's borders, were established to recognize the largest ethnic minorities within the RFSFR. Although this administrative order reflected a descending level of importance and, in a very general way, distance from the capital, that order could be countermanded through the powerful personality of a local leader. Kremlin watchers knew that when you began to hear the name of some remote kray frequently, it was likely that a party leader from there was ascending the political ladder and gaining influence in Moscow. Nor were all krays unimportant by definition: the one located farthest from Moscow, Primorskiy Territory, was home to the huge Soviet naval base of Vladivostok, a strategic city closed to the outside world whose borders were controlled even more tightly than those of the country as a whole.

The Russian "federation," of course, was a federal state in name only. The RFSFR, like the Soviet Union as a whole, functioned as a centralized unitary state, and all essential power resided in Moscow. The rights of ethnic minorities, despite their "republics" on the map, were strictly limited, and during World War II minorities tended to be suspect as potential allies of the invading Germans. The story of what happened to the Muslim Chechens is among the worst: Stalin accused them of collaboration and in 1944 ordered the entire population loaded on trains and exiled to Central Asia. Tens of thousands died along the way, their bodies thrown from the railroad cars. Many who survived this horror then perished in the harsh and unfamiliar environment at their destination. A man named Shamil Basayev, whom we will meet later, claims to have lost 40 relatives in this genocide, a fateful as well as dreadful personal calamity. Pardoned by Stalin's successor and permitted to return to their homeland in 1957, the remaining Chechens never forgot what Russians did to them, and when the Soviet Union collapsed in 1991 they seized on the opportunity to declare their independence. This started a cycle of violence that has killed thousands more and continues to this day—not just in Chechnya but in Moscow and elsewhere.

When Russia emerged from the wreckage of the U.S.S.R. as a geographically redefined country embarked on a course toward democratic government and a true federal system, Chechnya was not its only problem. The new Russian administration, led by the redoubtable Boris Yeltsin, could not simply sweep away the structural legacy the Soviets had built; it was the only game in town. So the Russian leadership took stock of the Soviet map and began to modify it to facilitate what would be a difficult transition. Counting all the administrative entities (whatever their rank) in the Soviet system, Russia was endowed with 89 regions, including 21 republics (Fig. 11-4). The regions were given an equal voice in the government through elected representatives, but special status was given to the republics. Early on, before Chechnya became Russia's nemesis, several of these republics proclaimed their independence, autonomy, right to self-determination, and other separatist intentions (Kaiser, 1994). One of them was historic and still substantially Muslim Tatarstan, astride the Volga east of Moscow, whose president in 1992 refused to sign the Russian Federation Treaty. Before long Tatarstan was flying its own flag over its assembly building, had launched an airline, and was insisting on "equal partnership" with Russia rather than mere membership in the 89-unit federal framework. In the Far East, a recalcitrant governor of the Primorskiy Region, now centered on an open Vladivostok where the Soviet fleet lay rusting and smugglers brought in contraband by the boatload, refused to send Moscow's cut of the (legitimate) tax take to the capital, defying the parliament. But here and elsewhere in the Russian federation, common sense eventually prevailed and notions of sovereignty faded. Still, although Russia did not fall apart during the Yeltsin presidency, chaos and near anarchy reigned as regional governors ignored the administration's laws, friends and associates of the president bought state enterprises for a song, organized crime flourished in Russia and spread its tentacles abroad, the armed forces sank into disarray, corruption was rife, and public health and well-being suffered.

TROUBLE IN TRANSCAUCASIA

As the government's failures mounted, the situation in Transcaucasia deteriorated. When the new Russia emerged in 1991, the Chechens shared their republic with their Muslim neighbors, the Ingush, but strife between the two groups had to be headed off. The Russian parliament in 1992 decided to split the republic in two, Chechnya and Ingushetiya, hoping to isolate the more militant Chechens from their more compliant neighbors. The effect, however, was to redouble the determination of Chechnya's non-Russians (30 percent of the population, then about 2 million, was Russian) to take control of their republic and to declare independence. It was the start of a cycle of increasing violence and failing negotiations that exposed the weaknesses of Moscow's government, the failures of its armed forces, and the depth of Chechnyans' resentment toward their rulers.

Chechnya, about the size of a small New England state, has three regions: the Caucasus mountains of the south, which provide refuge for insurgents seeking independence; the plains north of the Terek River, where Russians, who make up about 30 percent of the population, have been farming for

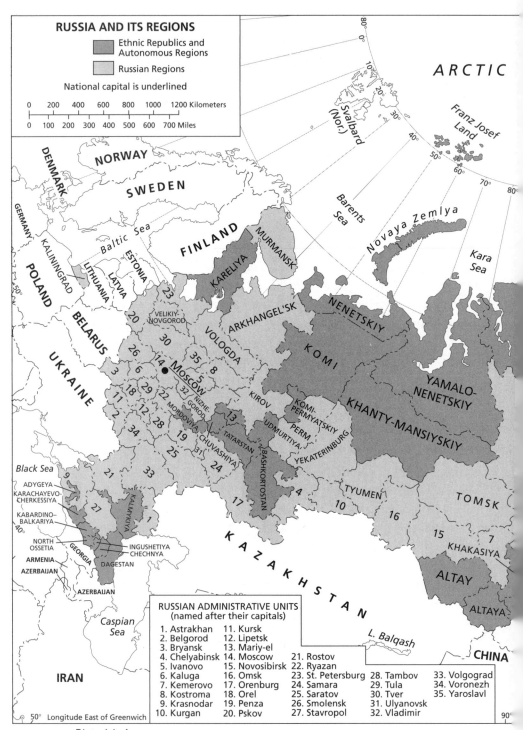

RUSSIA AND ITS REGIONS

Ethnic Republics and Autonomous Regions

Russian Regions

National capital is underlined

0   200   400   600   800   1000   1200 Kilometers
0   100   200   300   400   500   600   700 Miles

ARCTIC

DENMARK
NORWAY
SWEDEN
FINLAND
GERMANY
POLAND
KALININGRAD
LITHUANIA
LATVIA
ESTONIA
BELARUS
UKRAINE

Svalbard (Nor.)
Barents Sea
Franz Josef Land
Novaya Zemlya
Kara Sea
Baltic Sea

KARELIYA
MURMANSK
ARKHANGEL'SK
NENETSKIY
KOMI
YAMALO-NENETSKIY
KHANTY-MANSIYSKIY

VELIKIY-NOVGOROD
VOLOGDA
Moscow
KIROV
KOMI-PERMYATSKIY
PERM
UDMURTIYA
YEKATERINBURG
NIZHE.
GOROD.
MORDOVIA
CHUVASHIYA
TATARSTAN
BASHKORTOSTAN
TYUMEN
TOMSK

Black Sea
ADYGEYA
KARACHAYEVO-CHERKESSIYA
KABARDINO-BALKARIYA
NORTH OSSETIA
GEORGIA
INGUSHETIYA
CHECHNYA
DAGESTAN
ARMENIA
AZERBAIJAN
KALMYKIYA

KAZAKHSTAN
KHAKASIYA
ALTAY
ALTAYA

Caspian Sea
IRAN
L. Balqash
CHINA

RUSSIAN ADMINISTRATIVE UNITS
(named after their capitals)

1. Astrakhan
2. Belgorod
3. Bryansk
4. Chelyabinsk
5. Ivanovo
6. Kaluga
7. Kemerovo
8. Kostroma
9. Krasnodar
10. Kurgan
11. Kursk
12. Lipetsk
13. Mariy-el
14. Moscow
15. Novosibirsk
16. Omsk
17. Orenburg
18. Orel
19. Penza
20. Pskov
21. Rostov
22. Ryazan
23. St. Petersburg
24. Samara
25. Saratov
26. Smolensk
27. Stavropol
28. Tambov
29. Tula
30. Tver
31. Ulyanovsk
32. Vladimir
33. Volgograd
34. Voronezh
35. Yaroslavl

50° Longitude East of Greenwich

Fig. 11-4

Fig. 11-5

generations; and, between these, the urban-industrial middle zone where the capital, Groznyy, other towns, and the territory's oil installations are located (Fig. 11-5). Power over this middle zone means control over the territory, and here most of the conflicts that began in 1994 have been fought as the Russian army, at times exceeding 80,000 soldiers, fought rebels in a war conjuring up memories of Afghanistan, or perhaps Iraq. The capital was totally devastated, other towns were severely damaged, and negotiations after stalemates soon were followed by renewed fighting, with high casualties on all sides. Meanwhile what had been a nationalist campaign for independence turned into a wider war involving Islamic causes as fighters and funds from Afghanistan to Saudi Arabia arrived to support the Muslim Chechens (Tishkov, 2004).

However, Chechnya is only one of a tier of seven ethnic republics along the northern flank of the Caucasus mountains, from Dagestan in the east to Adygeya in the west (Fig. 11-2). Each one of these republics presents its own particular set of cultural, political, and economic problems in this fractious region: Dagestan's 2 million people are divided into some 30 ethnic groups; mainly Muslim Ingushetiya is divided between pro-Russian and pro-Chechen supporters; North Ossetians, generally pro-Russian, want union with their South Ossetian neighbors across the border with Georgia; the Muslim Balkars in Kabardino-Balkariya have memories of Russian mistreatment during World War II; many Muslim Karachay in the Karachayevo-Cherkessiya Republic were forcibly exiled during that war as well; and only

Adygeya is without latent ethnic conflicts. Sympathies for the Chechen cause extend well beyond Chechnya's borders, even after terrorists began to extend the conflict into this region and beyond.

This terrorist campaign, which soon reached Moscow itself, did incalculable damage to the Russian state and its political and economic hopes. Shamil Basayev, the Chechen who lost so many family members during the Chechens' Central Asian exile, was in Moscow supporting President Boris Yeltsin during the 1990s, seeing the president as Chechnya's only hope for autonomy. But when that hope was dashed with the 1994 intervention by Russian forces, Basayev was ready to play his role. In 1995, after Russian attacks leveled his family's home and killed 11 of his relatives including his wife, two daughters, and a brother, Basayev became Russia's Usama bin Laden. He began a terrorist campaign that has continued for more than a decade, unsettling not only the region around Chechnya but also Moscow itself.

Basayev's first high-profile action occurred a month after his family's demise, an attack on a hospital in neighboring Dagestan in which his fighters took 1,500 hostages of whom more than 100 were killed. In 1996 his forces drove the Russian army out of Groznyy and in effect achieved the autonomy Moscow had denied him, but in the "presidential" election that followed he received less than a quarter of the vote. Although he was included in the government, Chechnya in effect was a failed state by then, and Islamic jihadists and Arab funds were flowing in. Basayev became prime minister of the Chechen-controlled part of Chechnya they referred to as the Republic of Ichkeria. With Russia in political disarray and Boris Yeltsin about to relinquish his presidency, Basayev's terrorists mounted another raid into Dagestan and bombed two apartment buildings in Moscow. When the new president, Vladimir Putin, ordered Russian forces back into Chechnya in late 1999 for what was to be the second war for control, he had the almost universal support of Russians. The Chechen regime they attacked was recognized by only one other government: Afghanistan's Taliban. By midwinter 2000, Chechen forces had been driven out of devastated Groznyy and into their mountain hideouts, and from then on Basayev was reduced to planning and ordering a series of terrorist strikes. Perhaps the most dramatic was the takeover of a crowded theater during a performance in Moscow in October 2002 by a group of 41 Chechens and their allies, leading to a prolonged standoff, a bungled rescue effort, and the deaths of 130 theatergoers.

Consider the impact on Russia of such actions in just the year 2004: in February, a bomb in the Moscow subway killed 41 and injured more than 100. In May, a bomb planted under a row of seats in a Groznyy stadium killed the new Moscow-approved president, Akhmad Kadyrov. In June, an assault on police stations in neighboring Ingushetiya killed nearly 100. In August, a raid on police installations in Groznyy killed more than 50. In September, a team of terrorists took more than 1,000 children, parents, and teachers hostage in a school in Beslan, North Ossetia, resulting in the deaths of at least 370 children and adults. Also in September, suicide bombers simultaneously blew up two airliners flying from Moscow's airport, killing 90. With no prospect of changing the course of history through such actions, the Chechens, having lost their republic and seen many of their allies depart to fight the infidels in Iraq, now take opportunistic revenge and wait for a time when their cause will return to center stage.

DEMOGRAPHIC DISASTER

Russia's political, economic, and strategic struggles continue against a background of social problems so severe that they are routinely described by demographers as disastrous (Demko, 1998). In 1991, when reconstituted Russia emerged from the disintegrated Soviet Union, its population was approximately 148 million. By the beginning of 2005, it had declined to just over 143 million—even though several million ethnic Russians had immigrated during this period from the former Soviet republics. Geographers who study population issues calculate that, since the end of communism, Russia has seen about 10 million more deaths than births. Such population decrease usually accompanies a lengthy, major war or massive emigration. But in Russia, neither war nor an outflow of people is to blame. Rather, the situation signals severe social dislocation.

It should not be surprising that Russia's birth rate dropped markedly during and after the breakup of the Soviet Union, as uncertainty tends to cause families to have fewer children. Although the birth rate has more or less stabilized just below 9 per thousand, it is the death rate that has surged to more than 16 per thousand, causing an annual loss of population of over 7 percent or nearly 1 million annually. Only net immigration slowed a decline so severe that many demographers refer to it as catastrophic.

What is causing this calamity? The birth rate is held down by widespread abortion practices and by sexually transmitted diseases, but the death rate, especially among males, reveals the real trouble: rampant diseases such as tuberculosis, heart disease, and underreported AIDS, endemic alcoholism that is linked to these diseases and is also part of a culture of excess in which vodka and more recently beer play a major role, heavy smoking, especially among young males, traffic and industrial accidents, suicide, and murder. On average, a Russian male is nine times more likely to die a violent or ac-

RUSSIA: THE NEW FEDERAL DISTRICTS
PROCLAIMED MAY 13, 2000 AND THEIR CAPITALS

Fig. 11-6

cidental death than his European Union counterpart. Male life expectancy in Russia has declined from 71 in 1991 to 59 in 2004 (female life expectancy has also dropped, but much less, to 72). Fewer than half of today's Russian teenage males will reach 60.

What will happen to Russia if this continues? The Russian Parliament, the Duma, has been pressing the fast-growing beer industry to limit its advertising and has tried to legislate against beer consumption in the streets, but the culture of vodka consumption is so entrenched that this campaign is unlikely to have the desired effect. Yet if the rate of population loss continues, Russia may have a mere 100 million inhabitants by 2050, possibly fewer, and given its vast territory, this may so weaken the state as to make it unsustainable. Already, the Far East has lost 17 percent of its population since 1991, the South 12 percent, the Northwest, more than 8 percent and Siberia nearly 5 percent (Fig. 11-6). The Far East, with an area the size of the contiguous United States, now has a mere 6.7 million inhabitants, labor shortages are common, and development is mostly stalled (Thornton and Ziegler, 2002). The only solution seems to be a large influx of immigrants, and they are ready to come: North Koreans, Chinese, and others. But whether the government in Moscow will approve the immigration of as many as 250,000 East Asian immigrants is another matter. The Russian presence in the Far East is already tenuous, and the arrival of large numbers of Koreans and Chinese might create new social problems even as it begins to solve economic ones. Russia is in demographic trouble, and the way out is not in sight.

NEW ERA, OLD PROBLEMS

In 2000, Russia achieved a feat that a decade earlier was unimaginable: a democratic transition in the president's office. Vladimir Putin succeeded Boris Yeltsin, who for most of the decade held the office. Younger, more energetic, and determined to set Russia on a new course, he took on the oligarchs who had been favored by his predecessor, revived the economy, and began to reform the armed forces. He made it clear in public statements that he wanted Russia to acquire a stable political system, become a world-class economic power, and regain the international respect and strength it had lost during the nineties. He also proclaimed an adherence to the rule of law, and used this principle to corral several of President Yeltsin's favored oligarchs and put them in jail. But by rule of law, President Putin seemed to mean his own arbitrary power rather than a set of regulations passed by the Duma, the lower house of parliament.

Russia has a history of authoritarian government that goes back centuries, and the decade of President Yeltsin's chaotic and warped versions of democracy and capitalism made many Russians yearn for a stronger leader, even the old autocratic kind (Trenin, 2002). To a certain degree, this is what they got, and the ongoing conflict in Chechnya and its terrorist extension did much to create the opportunity. Immediately after the Beslan atrocity, President Putin said that "democracy does not result in stability, but rather instability . . . it does not unify, but rather divides" (Myers, 2004). Ethnic and religious tensions in culturally divided areas can only be controlled "with an iron hand from above." Most ethnic Russians appeared to agree with him, and many seemed willing to yield personal freedoms to accomplish it. Living in fear makes people worry less about individual rights and freedoms, and skillful exploitation of such an atmosphere can allow political leaders to concentrate power. President Putin approvingly pointed to the absence of Chechnya-like conflicts during the Soviet period, when such strife was "harshly suppressed by the governing ideology." Seizing the opportunity, the president began taking control of still-independent television, radio, and other media, and started manipulating regional and local elections.

The strategy to achieve more effective control over Russia's 89 politically still unpredictable regions was foreshadowed some years ago when the first Putin administration divided the country into seven new administrative units—not to enhance their influence in Moscow but to expand Moscow's authority over them (Fig. 11-6 displays these units). Each of these seven "federal" administrative districts has its capital city, which became the conduit for Moscow's "guidance," as the official plan put it. Then, in 2004, the other shoe dropped. President Putin introduced legislation to reverse the fundamental democratic right of representation in the regions, enshrined in the Constitution, by ending the regions' authority to elect its own governors. Henceforth, these governors would be appointed by the president, not elected by the people in the regions. Earlier, the governors had already lost their membership in the Federation Council, the upper house of parliament. Coupled with this move, the president announced plans to change the electoral system for seats in the Duma, so that party appointees, not independent legislators, will fill those seats. All this was approved in the Duma, where parties loyal to the president have a huge majority; it was even supported by some of the elected governors themselves. It was, in the words of a Russian colleague paraphrasing China's line on economics, a retreat to "democracy with a Russian face." Western governments, however, expressed strong reservations. In Bratislava, Slovakia during his visit to Europe in February 2005, President George W. Bush conveyed to President Putin American concerns over Russia's apparent drift toward authoritarianism in a private meeting followed by an awkward press conference that saw the two leaders talking past each other rather than with each other. President Putin reiterated his doubts about democracy; he cited America's Electoral College as evidence that American democracy has its own contradictions. The election of regional leaders in Russia, he argued, was no less democratic. As to the restrictions imposed on Russia's media, this, too, had precedents in other not-so-democratic "democracies." While the two presidents emphasized areas of agreement, including control over nuclear arms and the War on Terrorism, the issue of Russia's apparent course toward greater authoritarianism clearly eroded what had been a budding personal relationship.

RUSSIA AND THE WORLD TODAY

Russia may not have the capacity to recapture the position of global power once held by its Soviet predecessor, but this remains an important country. The status and operational condition of its nuclear arsenal is uncertain, but Russia remains a nuclear power. Russia's armed forces are in disarray, but reform is under way. Russia continues to wield influence in several members of the otherwise defunct Commonwealth of Independent States (CIS), an association of Soviet Socialist Republics formed to succeed the U.S.S.R. Russia maintains military bases and installations in 9 of its former 14 Soviet partners. Russia's large energy resources are already crucial to Europe and increasingly so to China and Japan. And Russia has taken independent, sometimes obstructionist positions in international strategic matters such as action in the former Yugoslavia, intervention in Iraq, and nuclear ambitions in Iran. For all his determined initiatives in fighting the War on Terrorism, President Putin declined to support the United States–led intervention in Iraq, taking an independent course when a different one might have yielded greater short-term benefits. Clearly the president wants his country to be a superpower again, an economic tiger in the making, and a political force to be reckoned with.

This chapter has enumerated some of the realities that cloud this vision, but there can be no doubt that Russia's energy position is favorable and highly lucrative. Indeed, it is a strategic asset as well. In 2005, China and Japan were vying for pipeline access to the oil field at Angarsk, not far from the southern end of Lake Baykal in the Siberia Region. The Chinese want to build a pipeline to a refinery at their Northeastern oil center of Daqing, whose long-productive reserves are dwindling. The Japanese want Angarsk's oil to flow to a terminal at Nakhodka on the Russian coast, from where it would be shipped to Japan across the East Sea. Moscow, aware of China's fast-rising demand for oil and of Japan's already substantial energy needs, is looking for the best deal.

With such economic advantage comes political clout. As noted earlier, the Russians have a long-standing territorial dispute with the Japanese over a group of four small islands off the northeast corner of Japan's northernmost large island, Hokkaido. These islands were occupied by the Soviets shortly before the end of the Second World War, and they never gave them back. Russia inherited them when the Soviet Union collapsed, and it did not return them either, despite continuing demands from Tokyo. These tiny Kurile islands are the reason why the Soviet Union, and now Russia, never signed a peace treaty with Japan ending their conflict. Over the years, various attempts to settle the issue have failed, despite generous offers by the Japanese to fund development projects in the Russian Far East, help develop ports and infrastructure, and initiate joint ventures. About 50,000 Russians have settled there, and when the United Nations Convention on the Law of the Sea (UNCLOS) took effect, the islands acquired not only a 12-mile territorial sea but also a 200-mile maritime exclusive economic zone (EEZ), enhancing their value. As to the Japanese, for all their bluster they cannot afford to play hardball with a Russia whose energy resources they badly need. The Russians will be on their Kurile outposts for awhile yet.

At the opposite end of Eurasia, Russia's relationships with the West, and especially with the European Union, are more complicated. Russia was remarkably cooperative during NATO's expansion to its very borders, and that cooperation was rewarded with NATO's practice of inviting Russian representatives to take their places at the table with NATO government ministers during administrative and planning sessions; Russians also participate in other alliance deliberations. This once-unthinkable arrangement symbolizes the new position Russia has taken in Eurasia and the world: it is no longer the adversary for which NATO was forged in the first place.

But the relationship between the European Union (EU) and Russia is less satisfactory. Those who argue that an EU capable of considering the eventual admission of Turkey should be able to contemplate the ultimate expansion of the EU into Russia face powerful counterarguments. Once again, Russia is a casualty of Chechnya: EU members abhor the brutality with which Russia has pursued its military objectives in the republic, its manipulation of elections, and its discrimination against minorities in the federation itself, justified as part of the necessary war on terrorism. EU governments see deterioration, not progress, in Russia's democratic reforms and human-rights practices, notably in the government's selective pursuit of "oligarchs" who enriched themselves during the chaotic Yeltsin presidency. Russia is the obvious state to encourage and support free, open, and fair elections in its CIS neighbors Ukraine and Belarus, but instead Moscow has been silent about Aleksandr Lukashenko's authoritarian actions in Belarus and meddled inappropriately in Ukraine's 2004 elections. Further, the Europeans want Russia, which has had a Partnership and Cooperation Agreement with the pre-expansion EU, to approve to a similar agreement with the new, 25-member Union, which now touches its very borders.

The Russians, on their part, have their demands too. Unlike NATO, they are on the sidelines when it comes to EU decision making, and they want a seat at the table in Brussels as well. They want the EU to buy more Russian oil and gas, aware that the resulting dependence will enhance their negotiating position. The Russians want trade concessions from the EU countries, because the new European map will close nearby (or common) borders against goods ranging from steel to farm produce; in addition they want the EU to agree to visa-free travel between Russia and the EU. They further want the EU to agree to give the Russian language official status in EU countries where large Russian-speaking minorities continue to reside, and an end to anti-Russian discrimination there. And always on the table is Russia's demand for free transit for commercial and military traffic between the exclave of Kaliningrad and Russia (a look at the map shows that such free transit, through Poland or Lithuania, will only reach Belarus, not Russia itself. Apparently no concern exists over transit through Belarus).

This enumeration of obstacles contains both intractable and far-reaching issues as well as comparatively minor ones. In Russia itself, a large and vocal segment of the parliament is in no mood to contemplate negotiations for EU

admission even if the EU proposed this; rather, they see Russia as a counterweight to the EU, with its own economic and political sphere including
Ukraine, Moldova, Belarus, Georgia, Azerbaijan, Armenia, and even some
of the Central Asian republics. After all, Russia still has a population much
larger than any EU member, is territorially more than three times as large as
the European Union, has a distinct culture, is ethnically more complex than
any neighbor near or far, and has a history of empire unlike any European
country. An Atlantic-to-Pacific European Union would constitute an ultimate confirmation of the goals set in Rome half a century ago, but it may
take that long again to accomplish.

12

HOPE FOR AFRICA?

Informal but persuasive surveys indicate that, among geographic realms competing for American attention, Africa ranks dead last—the "real" Africa, that is, the Africa lying south of the vast Sahara, the Africa of the defiant Ashante and the mysterious builders of Zimbabwe, of the powerful Zulu Empire and the stone city of Zanzibar, of the bustling markets of Dakar, the oil platforms of the Niger Delta, and the gold mines of the Witwatersrand. It is the Africa of the vast Congo Basin and the great Kilimanjaro, of Victoria Falls and Table Mountain, of the Great Lakes and the Kalahari. There is nothing like it in the world, its diverse peoples a kaleidoscope of cultures, its endangered primates a mirror of humanity, its dwindling wildlife populations a fading link with the early Tertiary.

It is also *terra incognita* to Americans more than any other part of the world. When Africa does gain America's attention, it tends to happen because of civil wars, health crises, natural disasters, or terrorist attacks, and only rarely because of the kinds of positive developments that occasionally emerge from other parts of the world. When the murderous dictator Abacha ruled Nigeria, he and his excesses were regular fodder for United States newspapers, but when the country achieved a remarkable, generally peaceful transition to democracy and President Obasanjo was elected, his tribulations in Africa's most populous and religiously divided country gained far less attention. South Africa's dramatic transition from apartheid to democracy generated a brief surge of interest and endowed President Mandela with celebrity status, but how much attention are the United States media giving today to what is by many measures Africa's most important country?

United States leaders do from time to time signal momentary awareness of Africa's plight and use it to make high-profile forays to states deemed deserving, as did President Clinton during his second term (with little or no outcome despite much emotional rhetoric), or highly publicized commitments to help solve Africa's problems, as did President George W. Bush during his first term with a $15 billion fund to combat AIDS (but also to promote his constituency's views on family planning). When Bush's first secretary of state, Colin Powell, was appointed he raised the hopes of many when he stated that Africa would rank among his highest priorities during his term in office. The events of September 11, 2001, and their aftermath ended that initiative before it could begin.

In the public eye, therefore, Africa south of the Sahara, the ancestral home of hominids and the cradle of humanity, the setting of our first communities and the scene of our first cultures, the place where we made our first tools and spoke our first words, the theater of our first artistic expressions and the landscape whose Pleistocene variability would propel our ancestors into Eurasia and the world—that Africa is largely forgotten. Originally we are all Africans, and to know ourselves better we should reconnect with the source. Our territorial and environmental imperatives began as African experiences, and some modern people, upon seeing Africa for the first time, experience an epiphany that may have scientific implications. The biologist Edward O. Wilson reports asking his students (and others) to draw what they conceive of as their ideal natural landscape; when he distilled a consensus out of thousands of such drawings that consensus resembled the East African savanna (Wilson, 1995). We may have left Africa, but Africa has not left us.

It is therefore appropriate to end our geographic journey in Africa, because this is where the human saga began and where many, indeed most, of the world's problems converge today. Presidents Vladimir Putin of Russia and Thabo Mbeki of South Africa have this in common: they both recognize the effect of extreme poverty on human behavior. Putin, following a visit to the terrorist-targeted school in Beslan, railed against the "excesses" of democracy but also observed that people who are so poor that they have nothing to lose cannot be expected to accept their fate indefinitely. Mbeki attributed South Africa's AIDS epidemic to poverty rather than the virus that causes it; he was medically wrong (at great cost to his nation) but socially right. If the world is becoming a global village, Africa is its poorest neighborhood, and when a town's poorest neighborhood gets help, the whole community benefits. If the European Union can siphon funds from its wealthiest members to assist its less prosperous ones, then surely the world can find ways to help Africa's neediest peoples.

African countries, unfortunately, rank high on the corruption ladder—none is a Finland or a New Zealand, or even a Chile. So the argument will be made that funds allocated are dollars wasted. But, as we will see, there are other ways to help Africa. And there is ample proof all over Africa that, given a chance, farmers and furniture makers, diamond cutters and doctors are as good as any in the world. It is just that they need a chance to prove

their capacities in a world that has, in more ways than one, turned its back
on them and their region. In late 2004 the economist Jeffrey Sachs tried to
rally worldwide public support for a massive aid program for Subsaharan
Africa through a series of public lectures as well as articles in the *Economist*
and other journals, proposing a global "Marshall Plan for Africa" that would
begin to address the most urgent of Africa's ills. He proposed ways to target
such aid, involving tens of billions of dollars contributed, like the European
Union's subsidy funds, by the richest nations of the world, directly to the
neediest sectors of African economies. It was time, he argued, to invest in a
better world by investing in Africa. The expense would be less than the annual cost of the war in Iraq.

Africa suffers from a series of misfortunes, several of them geographic, so
severe that they have, in combination and in perpetuity, put Africans at an
incalculable disadvantage. This is not over—AIDS is only the latest manifestation of just one of them—and it has reached the point where what benefits
much of the rest of the world can actually hurt Africa's chances to catch up.
Shortly after the remarkable transition to democratic government in South
Africa, I happened to be in Australia where I heard President Mandela's Af-
rikaner partner in this remarkable process, F. W. de Klerk, present an address, unreported in the American media, in which he argued that what the
world needed was an Indian Ocean version of the Pacific Rim phenomenon:
an Indian Ocean Rim anchored by South Africa, Australia, Thailand, and
Western India that would transform the economic geography of the Southern Hemisphere and "carry Africa into the twenty-first century." But in all
of Africa, only South Africa and possibly southern Moçambique might see
some benefit from that vision. Australia and Southeast Asia were already
in the Japanese-Chinese economic sphere. Unlike their putative partners,
African countries tend to have little economic clout or political influence
in the wider world. It will take more than free-market economics (not free
for most African producers, anyway) to overcome the cumulative setbacks
chronicled here.

EIGHT FORMATIVE DISASTERS

Africa's condition is a matter of statistical record. By any combination of
criteria, from income levels to diets, infant mortality to life expectancy,
health to literacy, Subsaharan Africa is the world's neediest geographic
realm. When the Holocene epoch opened, Africa had for tens of thousands
of years witnessed the trials and achievements of emerging humanity. But
today, Africa and Africans, most of them descended from the first humans

to walk on this Earth, suffer the effects of a combination of long-term disasters, beginning thousands of years ago, unlike those befalling any other corner of the globe. If non-Africans avert their eyes from the realm, it is because they see no hope and fear that the future will be still worse, not better. Those who could make a difference, including the so-called international community, meaning the powerful and richer states, leave Africa mired in its misery, sometimes blaming the victims for their problems. And it is not a matter of minimal subsidies; it is a matter of giving Africans a fair shake on world markets for products ranging from cacao to cotton. Champions of free trade deny African farmers opportunities on world markets through subsidies that, if ended, would send hundreds of billions of dollars to African cultivators.

But Africa is as powerless in the new unipolar world order as it was during the old bipolar world order, and it is a condition that has arisen from an incomparable series of misfortunes over the past 10,000 years, a combination of disasters that distinguished Subsaharan Africa from all other world geographic realms. Some other realms have suffered individual calamities that may exceed what happened to and in Africa, but none has experienced the coalescence of catastrophes that put Africa at the disadvantage it confronts today, and is reflected in indices ranging from life expectancies to incomes, food availability to infant mortality, and from disease incidence to education. Here are the sources of Africa's plight:

Climate Change

When the planet emerged from the extreme cold of the late Wisconsinan Glaciation starting about 18,000 years ago and warmed to the beginning of the Holocene epoch after the brief reversal to bitter cold during the Younger Dryas starting 12,000 years ago, the climatic and vegetative zones that lay compressed between periglacial and tropical latitudes began to shift poleward. That shift had momentous consequences for Africa, because the freezing up of Europe had pushed moist and moderate conditions into what is today the Sahara, where forests stood and streams flowed. Equatorial Africa was significantly cooler and equatorial rainforests had yielded to savannas, but now the warming brought rainforest expansion. As time went on and Egyptian civilization arose in the lower Nile River basin, the Sahara region became increasingly desiccated by 5,000 BP, creating a vast natural barrier between Mediterranean and tropical Africa. The Nile route continued to link Egyptian and African peoples, but the exchange of innovations between north and south was inhibited by environmental transformation across the entire northern bulk of the continent with consequences that would permanently isolate "Subsaharan" Africa. Eventually Egypt and Ethiopia, Morocco and Ghana would be worlds apart, separated by thousands of miles of inhospitable rock and sand, linked only by the most tenuous overland (in the west) and overwater (in the east) connections. Imagine instead an adequately watered, peopled, productive Africa extending to the very shores of the Mediterranean across from Europe, a Pan-African highway and road system linking Cairo to Cape Town, Algiers to Accra, Casablanca to Kinshasa.

The formidable Sahara barrier, its Sahel margin pulsating cyclically into what remains of its fragile adjoining ecologies, ensured that Africa would be regionally fractured and vulnerable to demographic crises, and that Africa would be "North" and "Subsaharan."

Ecological Impact

Not only did the Holocene's warming shift biomes from lower to higher latitudes; it also intensified equatorial climates in low-latitude and low-altitude Africa within and well beyond the margins of the Congo basin. That intensification produced higher temperatures, higher humidity, luxuriant rainforest growth and expansion, exuberant animal life from forest to savanna, the formation and filling of lakes and swamps, and, as it turns out, the final natural enlargement of the habitat of the great apes that had survived numerous previous environmental swings of the Pleistocene.

For Subsaharan Africa's human population, this global-warming episode had dire consequences. *Homo sapiens* had begun to emigrate from Africa via the southern end of the Red Sea as long as 90,000 to 85,000 years ago (an earlier attempt across the Sinai Peninsula had failed because of the sudden onset of the Wisconsinan Glaciation) but those who remained on the continent faced an environmental challenge that was to put them at a critical disadvantage when plant and animal domestication began elsewhere (Oppenheimer, 2003). While there is little doubt that African farmers did domesticate grains in West Africa as well as the Horn, animal domestication was another matter. Proliferating African wildlife was a threat to human existence, not an opportunity. While oxen, goats, and other animals were being domesticated in the Middle East and elsewhere in Eurasia, Africa's fauna defied domestication. As Jared Diamond reports, the only animal known to have been suitable for domestication in Subsaharan Africa was the guinea fowl (Diamond, 1997).

The heat and humidity of Africa's equatorial lowlands generated still another threat to African well-being: a host of diseases ranging from malaria to bilharzia and from yellow fever to sleeping sickness. Throughout the Holocene, the incidence of diseases rose, and even today the great majority of the approximately 1 million annual new victims of malaria are Africans. The equatorial environment nurtures countless vectors, from mosquitoes (malaria, yellow fever) and flies (sleeping sickness, river blindness) to worms and snails (the latter transmitting bilharzia), and people living in close proximity to wild animals, as well as consuming them, run further risk of transmission. Africa's problems in this regard are not just those of the distant past: sleeping sickness appears to have originated in West Africa as recently as the fourteenth century, and AIDS, Africa's latest scourge, began in the second half of the twentieth century. In 2004 Subsaharan Africa had 70 percent of the world's recorded AIDS cases and 75 percent of its deaths. Life expectancies were falling throughout the realm and projections indicated that AIDS would leave 20 million African orphans by 2010 (Altman, 2002). But in long-term perspective, AIDS is only the latest environment-related health calamity to strike Africa. The health of the general population has been poorer in Africa than anywhere else in the world for countless generations; remedies against African diseases have not been sought nearly as aggressively as has been the case for other areas of the world; and Africans tend to be too poor to be able to afford those medicines that do exist. For Subsaharan Africa's peoples, that combination of risk and neglect has been disastrous (Best and de Blij, 1977; Aryeetey-Attoh, 1997).

Divisive Islam

Surely the arrival of a great universal religion cannot be deemed disastrous for any realm or region of the world? In Africa's case, the Muslim faith came to unite, to sweep away local belief systems, to focus Africans' gaze on Mecca—but by its partial, incomplete penetration it also had the effect of dividing Africa along the Islamic Front referred to in chapter 8. That line, in effect a narrow transition zone, extends from Guinea in West Africa to Ethiopia and the Somalia-Kenya border in East Africa. It cuts across the middle of Ivory Coast, divides Nigeria nearly in half, fragments Chad and Sudan, and cuts across the Ogaden in Ethiopia. Its social and political implications are momentous; it has precipitated wars costing millions of lives between Arab and African and between Arabized (Islamized) African and Christian and animist African.

Islam arrived in West Africa by caravan and in East Africa by boat, and in West Africa its proselytizers converted the kings of the great and stable indigenous states that prospered from the trade between peoples of the coastal forests and those of the drying interior. With the successful Islamization of Andalusia in full swing, the Muslims turned their attention to Saharan Africa, and by the fourteenth century they had done in the Saharan region what they had accomplished in Arabia seven centuries earlier. West African rulers made Islam the state religion, and huge, rich pilgrimages made their way along the savanna corridor from the states flanking the Niger River to the holy centers across the Red Sea. These pilgrimages, involving tens of thousands of people every year, played a major role in changing the ethnic map of the region: many stayed behind, on the way to or from Mecca, implanting their cultures from Nigeria to Sudan.

But Islam's southward march was halted by African resistance and by European-Christian intervention, and the Islamic Front came to reinforce Africa's division between Northern and Subsaharan. Now it was not just a desert but also a faith that reinforced the continent's partition, a further barrier between two realms and, ultimately, a source of devastating conflict.

The Depopulating Slave Trade

Putting people in bondage and working them to death was a worldwide practice (the word "slave" comes from the ethnic term *Slav*, Eastern Europeans enslaved by Muslim Turks). The ancient Greeks and Romans had slaves and the powerful enslaved the weak or the defeated in many, though not all, cultures. But nothing in the world compared to what happened to Africa when Europeans arrived not just to enslave Africans, but to capture, transport, and sell them overseas. True, Arabs had been carrying on a slave trade from East Africa for centuries before the Europeans organized their version of it, mainly from West Africa but also from other areas. It was the dimension, the sheer volume of the European slave trade that dwarfed all other forms of the practice (Curtin, 1969).

Much has been written about the numbers, New World destinations, and dreadful treatment of those who were taken in bondage from their African abodes and who survived the unimaginable Transatlantic voyage. Estimates of the dimensions of the traffic between 1700 and 1810 range from 12 million to more than double that figure; the world will never know. Still less is known about the consequences of the slave raids in Africa itself. Subsaharan Africa's population at the beginning of the eighteenth century may not have been much more than 90 million (the estimate for the global population in 1700 is 650 million), so that if the number of forced migrants was 15 million, this means that one-sixth of Africa's population was taken to the Americas (Fig. 12-1).

Anthropologists reckon that the impact of this terrible, catastrophic event can still be read in Africa's cultural landscapes. Not only were entire areas depopulated by the slave raiders as well as through the fighting that attended the campaign, but the price put on human heads set African against African and

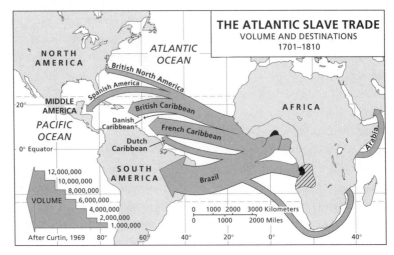

Fig. 12-1

rekindled ethnic animosities. Children were orphaned or abandoned by the hundreds of thousands, crops lay unharvested, villages stood deserted if they were not burned. West Africa's forest-to-desert trade collapsed, the Islamized interior states broke up, and everywhere the social order disintegrated. Columns of chained Africans marched from ever-deeper inland "sources" to the coastal embarkation stations, never to see their homeland again.

By itself, a disaster of such magnitude would have lasting consequences for any cultural realm, but in Africa the slave trade was only one of a series. When the slave trade ended and the Europeans, now taking the moral high ground, closed the last of the Arab slave markets in East Africa, the damage done to Africa's social fabric was incalculable. But the foreign powers that had conducted it were still there, pursuing other goals.

Colonialism

Europeans had been testing African waters long before the slave trade started, and they would remain as occupiers long after it ended. European trading posts were founded along the West African coast in the 1500s to outflank the commerce that was still linking coastal states, savanna empires, and Saharan caravans, including the sizeable and lucrative trade in gold. Portuguese vessels rounded the Cape of Good Hope and entered the Indian Ocean, the Dutch established a supply station at Cape Town in support of their East Indies traffic, and the British and French, followed later by the Belgians, Germans, and Italians, staked their claims to African empire.

Most of this activity, though, was concentrated in Africa's coastal areas. Well into the nineteenth century the interior remained the sphere of explorers and missionaries, raiders and occasional traders. When territorial issues began to arise and the fateful Berlin Conference was convened late in 1884 to settle conflicting claims and to carve Africa up among the competing colonial powers, much of the continent's vast and remote interior was barely known to the contestants. Boundaries were drawn to codify spheres of influence, but in many areas there was not enough information to gauge how these colonial borders would affect local peoples. As a result, some of those boundaries split unified ethnic groups into different jurisdictions; others combined peoples with historic animosities. Still others disastrously interfered with seasonal migrations, depriving pastoralists and their herds of water and forage when Berlin's borders were imposed on the ground (Newman, 1995).

Colonialization meant exploitation, and exploitation required terror. In the current era of terrorism it is well to remember that a century ago African peoples were terrorized by European colonists on a "civilizing mission" that subjugated an entire culture realm, a mission that was carried out by a small minority of invaders whose principal means of maintaining control was fear. The Europeans' shared objective was the exploitation of Subsaharan Africa's natural resources, and in this pursuit they established administrative head-quarters, transport routes, and ports, creating the beginnings of Africa's modern, disconnected infrastructure. To be sure, their methods differed: in general, British colonial practice was more benign than that of the French or the Portuguese, and the Germans and the Belgians were the most ruthless. The short-lived German occupation of four African territories was referred to in the European press as "colonization by the Mauser" because so many tens of thousands of Africans were killed by this efficient gun; the story of the Belgian King Leopold's "Congo Free State" ranks among the century's worst human calamities.

The world should reflect on what was done to the approximately 20 million inhabitants of the territory awarded to King Leopold during the Berlin Conference, because the aftermath still lingers in this blighted country: "King Leopold II . . . embarked on a campaign of ruthless exploitation. His enforcers mobilized almost the entire Congolese population to gather rubber, kill elephants for their ivory, and build public works to improve export routes. For failing to meet production quotas, entire communities were massacred. Killing and maiming became routine in a colony in which horror was the only common denominator. After the impact of the slave trade, King Leopold's reign of terror was Africa's worst demographic disaster. By the time it ended, after a growing outcry around the world, as many as 10 million Congolese had been murdered" (de Blij, 2004).

The destructive impact of imperial pursuits in Africa are well enough

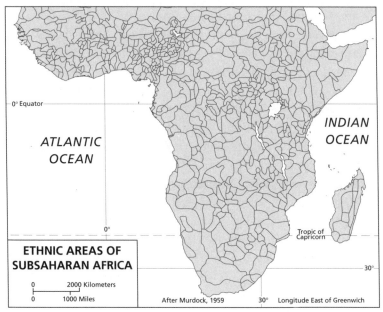

Fig. 12-2

known today not to require further elaboration, but colonialism had a more subtle yet equally enduring effect. Subsaharan Africa in colonial times was culturally a highly diverse geographic realm, as the famous "Murdock Map" of ethnic areas, published nearly a half century ago, underscores (Murdock, 1959). But those lines delimiting the traditional domains of Africa's ethnic groups are deceiving. True, as in all parts of the world, there are well-defined ethnic borders in Africa. More often than not, however, the lines on the Murdock Map represent transitions, zones where African peoples with different cultural traditions lived together (Fig. 12-2). Already it was, in many areas, difficult to tell representatives of one "tribe" from another. In present-day Rwanda and Burundi, for example, where sedentary Hutu farmers had been invaded by pastoral Tutsi, some Hutu had become landowners and herders in the Tutsi tradition and a process of cultural convergence was under way.

As John Reader recounts, one common practice among colonial powers with otherwise divergent policies was to classify all their African subjects on the basis of ethnicity. Whatever the objectives (to facilitate "indirect rule" in British dependencies, to organize forced cropping on Portuguese farms, to collect taxes everywhere), the policy had divisive effects that would long outlast the colonial era. It was retribalization that enhanced the status and power of some peoples while diminishing others; it assigned Africans with mixed backgrounds to often unfamiliar social status; and it laid the groundwork for conflict in postcolonial states (Reader, 1998). The policy reached its apogee in apartheid South Africa, where the variant was domestic, not European, colonialism and the results continue to plague the multiethnic nation.

It is tempting to speculate on the kind of Africa that would have emerged in the absence of colonial intervention and its associated boundary demarcation, land alienation, urbanization, and other consequences of what was in effect Subsaharan Africa's first experience with globalization. Africans had already proven their capacity for building long-lasting, multi-ethnic states; they had already built major cities and engaged in long-range, inter-regional trade (Newman, 1995). We will never know, but of this there is no doubt: the colonial episode was another in Africa's unparalleled series of formative disasters.

The Cold War

In the aftermath of World War II Africa's prospects appeared to improve. The colonial powers had fought among themselves and weakened their ability to hold overseas empires (France and the Netherlands in Southeast Asia, Britain in South Asia), and the superpower Soviet Union proclaimed itself a bulwark of anticolonialism. By the 1950s, African resistance had grown strong and negotiations were under way to facilitate the transfer of power from London and Paris to such capitals as Accra, Lagos, and Dakar. Rebellions and larger conflicts pitted Africans against Europeans in such territories as Kenya, Congo, and Angola, and white settlers held back the decolonization process in (then) Rhodesia, but within little more than two decades, starting with the Gold Coast (renamed Ghana) in 1957, Africa's colonial era ended and several dozen newly independent states appeared on the map (de Blij, 1991).

Inevitably, such rapid transition from colony to statehood produced crises and conflicts. Some colonial powers had done more than others in fields such as higher education and civil administration to prepare their dependencies for sovereignty. Britain left Ghana with a Westminster-model government and a strong economy and helped Nigeria deal with its regional schisms by developing a federal system. Numerous African university graduates from African as well as British universities gave these new administrations authority as well as authenticity. The Belgians, on the other hand, had done little to generate a college-educated cadre of Congolese although they knew that independence was inevitable, and the result was chaos. In addition, conflict quickly arose among nations in states that had held comparative power and privilege during the colonial period and other nations that saw, in independence, an opportunity to improve their position. Kenya's Kikuyu, Uganda's BaGanda, Nigeria's Yoruba, and many other prominent African nations were thus challenged by others within the state, so that politics soon tended to take on a "tribal" tone.

Such transition-engendered conflicts arose at the worst possible time for Africa, because many of them were magnified by the superpowers competing in the Cold War. The Soviet Union, itself a colonial empire, saw an opportunity to legitimize its anticolonial credentials by supporting those African groups that professed leftist leanings. The United States, which had vigorously supported African independence, had appointed a secretary of state for African affairs, and sent thousands of Peace Corps workers to Subsaharan Africa, nevertheless acted also to obstruct Soviet designs. The CIA was implicated in the murder of the first Congolese prime minister. From Ethiopia to Angola, proxy wars raged and hundreds of thousands of ordinary Africans died of war, starvation, and dislocation. Millions of land mines continue to take a daily toll and billions of dollars will be required just to bring infrastructure back to where it was four decades ago.

The Cold War caused the United States to turn a blind eye to the excesses of African rulers who could be trusted to oppose communism, including the Congo's Mobutu Sese Seko, Ethiopia's Haile Selassie, and others; along with the former colonial powers it tolerated murderous tyrants like Idi Amin (Uganda) and Jean-Bedel Bokassa (Central African "Empire"). With the Soviet Union encouraging the likes of Mengistu Haile Mariam (Ethiopia) and Agostinho Neto (Angola), millions of Africans had traded foreign imperial rulers for domestic ones under the sway of foreign powers. It was another damaging turn of events for a realm falling ever farther behind the rest of the world.

Globalization

Once again the world is changing dramatically, as it did during the era of colonialism and during the wars of the twentieth century. Today the changes come through a series of interrelated processes referred to as globalization, the breaking down of barriers to international trade, the stimulation of commerce, the revival of long-dormant economies, the stimulation of social, cultural, political, and other kinds of exchanges. In the rich and powerful countries of the world, globalization has enthusiastic support although small but vocal minorities oppose it. But in the less privileged parts of the world, where participation in the process is not easily achieved, globalization is seen by many as a threat, not as an opportunity.

Analyses of the worldwide impacts of globalization underscore the negative effects the process has on most countries in Subsaharan Africa. While it is evident that an openness to globalization in its various forms is advantageous to countries already comparatively wealthy or on a relatively rapid developmental path, the poorest countries, most of them in Africa, with weak economies propped up by unwise tariffs and other trade barriers, lag ever farther behind. Thus globalization is increasing, not decreasing, the gap between poor and well-off; statistical averages notwithstanding, the lot of the average African has not improved in the era of globalization. The privatization of some of the few state-held assets in these poorest countries, or the construction by a global corporation of a coastal casino, luxury hotel, or beach resort in Moçambique or Tanzania may suggest modernization and may be reflected in "development" data, but does nothing to change the fundamental condition of the state. The geographer Robert Stock has described the poverty-stricken condition of Subsaharan African cities, many of which are in worse condition today than they were at the end of the colonial era (Stock, 1995). Urban landscapes reflect Africa's remoteness from the forces of change that are transforming urban centers from Santiago to Seoul.

Is globalization a disaster for Africa on a par with the other seven chronicled here? It is yet another process of extra-African origins that has a deep and long-term impact on Africa's internal condition and external position. Globalization's meager penetrations in Africa (South Africa excepted) have stoked corruption and threatened fragile cultures and ecologies. Analysts sometimes observe that Africans have themselves to blame for the loss of opportunity associated with globalization's slow march in this realm, but globalization is a power play, and power on the global state is not an African asset. The cumulative consequences for Africa constitute a terrible misfortune.

The Failure of Leadership

During the 1960s, when Europe's African dependencies followed Ghana's lead and marched toward independence, African leaders garnered worldwide admiration. Many of these leaders had spearheaded independence movements long before the colonial powers began to contemplate a transfer of power, some led armed insurrections and many served time in prison. The exhortations of Ghana's Kwame Nkrumah inspired African nationalists everywhere; the detention of Kenya's Jomo Kenyatta during the Mau Mau uprising symbolized the struggle for all. When France offered a strings-attached sovereignty to its West and Equatorial African colonies and Guinea was the only prospective state to turn this down, Sekou Touré became a hero from Dakar to Durban. In the early years following independence these "fathers of the nations" appeared to prove eloquently what Africans were capable of, and what the colonists had mindlessly suppressed. Tanzania's Julius Nyerere translated Shakespeare into Swahili; Senegal's

Leopold Senghor was elected to the French National Academy of Sciences. African leaders spoke eloquently at United Nations conferences and meetings of nonaligned states during the Cold War.

But the route from father of the nation to usurper of the people turned out to be a short one. Military coups ended many brief experiments in democracy and ousted elected leaders, but elsewhere the legitimate successors to the nations' founders had fewer principles than their predecessors had. Even as murderous regimes such as those of Amin (Uganda), Bokassa (Central African "Empire"), Abacha (Nigeria), Mobutu (Congo), and Doe (Liberia) shocked the world with their excesses, other, more stable African states began to appear on lists of the world's most corrupt countries. Kenya, where Daniel arap Moi had succeeded Jomo Kenyatta, headed this depressing roster.

In Southern Africa, the long-imprisoned Nelson Mandela made possible that almost unimaginable, peaceful transition from apartheid to democracy, setting one glorious and final example of African leadership and statecraft in the liberation era. But in South Africa's neighbor, Zimbabwe, the revolutionary hero Robert Mugabe turned into another of Africa's destructive tyrants—in the process ruining one of the continent's most promising economies. When Nelson Mandela was democratically succeeded in South Africa by Thabo Mbeki, the new South African president found it difficult to express toward Zimbabwe the moral standards he inherited from the founder.

African leadership has failed in Africa, but the leaders of the "international community" have failed Africa too. Former European colonizers turned a blind eye to the misdeeds of the dictators with whose countries they did business; Cold War politics prevailed over principle when ruthless rulers took sides; and more recently the convulsions of the Congo and its eastern neighbors, costing more than 3 million lives, failed to stir the world into the massive intervention the region, by some measures, was surely owed.

In this first decade of the new century there are some brighter prospects: a return to democratic government in Nigeria and a democratic election in Kenya, peace and the start of reconstruction in Angola, stability and progress in Senegal and Ghana, and other signals of new era. But a half century of widespread failure has taken a toll that will long be imprinted on the African map.

Given the millennia-long concatenation of misfortunes just chronicled, what are Africa's prospects in the twenty-first century, and why should Africa reclaim the priority position in international affairs it briefly held during the period of decolonization? The consequences of a misgoverned, unstable, unhealthy, poverty-stricken Africa to the rest of the world are reason enough. Africa's well-being and global connectivity are important to the world, and especially important to the world's sole superpower of the day.

WHY AFRICA MATTERS

"Always something new out of Africa," exclaimed a Roman emperor 2,000 years ago, and so it is today. In the rainforest of the northeastern Congo, workers paid a pittance for their labors are digging from the ground a raw material used in the manufacture of cell phones. From uranium in the atomic age to oil in the fossil-fuel era, Africa has always had what it takes—for the rest of the world.

But concern for Africa's well-being should not focus on the relentless acquisition of its commodities. Africa's problems and the world's concerns coincide because the world is functionally shrinking, and when one of the neighborhoods of the "global village" suffers more than any other from a combination of maladies, the remedy benefits all. So assisting in the recovery of Subsaharan Africa is not mere altruism; it is a matter of self-interest for the rest of the world—and especially for the United States.

In the first place, social stability in Africa (as elsewhere) is a precondition for progress. Recent events in West and Equatorial Africa gave rise to fears, prominently expressed by Robert Kaplan, that the civil wars in Sierra Leone and Liberia would spread widely, precipitating a "regional anarchy" (Kaplan, 2000). Instead, a small contingent of British troops quelled the violence in Sierra Leone during the late 1990s and arrested the movement's leader, Ferdie Sankoh. And when American forces failed to intervene in Liberia an army of Nigerian peacekeepers drove the country's ruler, Charles Taylor, into exile in 2003 (he is held responsible for the deaths of more than 300,000 people by starting ethnic conflicts that spilled over from Liberia into Sierra Leone). In Equatorial Africa, what started as another in a series of Tutsi-Hutu conflicts in Rwanda spilled over into the eastern Congo and ended the rule of the feared Mobutu, but not before drawing in an assortment of regional profiteers who saw economic opportunity in the chaos. In 2004, leaders of the Hutu campaign were on trial in Tanzania and a new "coalition" regime in Kinshasa seemed to be making progress toward a lasting resolution.

The greatest danger to African stability lies along the "Islamic Front" referred to earlier (Fig. 9-2). Several postcolonial states are bisected by this cultural transition line with attendant risk, including Côte D'Ivoire (Ivory Coast), recently the scene of bitter conflict between southerners and northern Muslims, and Nigeria, where 12 of the federation's 36 States have adopted Islamic sharia laws (Fig. 12-3). The schism between north and south in Nigeria is fraught with danger, and cultural distinctions have grown sharper over the past decade as hundreds of thousands of southerners living in the north have moved away. The stability of Nigeria, cornerstone of Africa's historic cultural core, is indispensable to Africa's future, and the country's return to

Fig. 12-3

democracy should be acknowledged through substantial international assistance toward its infrastructure needs.

Islamic outposts as far away as Cape Town form a reminder of the susceptibility of Subsaharan Africa to Islamic proselytism, religious as well as political. The vulnerability of the poor, the frustrations of neocolonialism real and perceived, the association of Christian faiths with imperial histories, fears of globalization, and promise of empowerment give Islam's proponents opportunities in Africa that may change the ideological map of the world. To counter this prospect, Western involvement in Subsaharan Africa must grow exponentially, supporting democracy, building infrastructure, opening markets, and strengthening linkages.

Which leads us to the third consideration under the present rubric, the question of products and commodities. Western involvement in Africa should go far beyond the purchase of oil and minerals, and requires sacrifices from all wealthy countries, Western and non-Western. The matter of agricultural subsidies ranks high in this category. The majority of Africans today still make their living as farmers, subsistence as well as commercial, and for the latter, any unfair competition on "free" markets has devastating results. But that is exactly what African farmers face, from France to Japan to the United States. African farmers earn very low per-hour wages, and they are able to market their tea, cotton, cacao, and bananas at very low prices. But they cannot compete with farmers who receive state subsidies to plant, harvest, export, and market their produce. It is up to the rich countries to phase out these subsidies—if they really believe their own rhetoric about free trade—and to give African farmers their chance. A recent World Bank estimate indicates that African farming would benefit to the tune of more than $200 billion per year, more than 20 times the financial aid currently given to Africa by donor countries.

African farming is in a poor state, and to be sure there is more involved here than the policies of the richer countries. African governments themselves have failed to support their farmers in several ways: by keeping prices of staples artificially low, pleasing urban residents but costing farmers dearly, by making it difficult for farmers to secure loans, by investing heavily in ill-conceived industrial projects rather than the important agricultural sector of the economy, and by denying women, who do much of the actual work in the fields, credit opportunities to start and maintain their own farms or businesses. Add to all this the severe seasonal fluctuations in Africa's weather and the marginal productivity of much of Africa's soil, and it is clear why Africa's per-capita farm production has been declining for more than two decades.

If there is little incentive in the richer world to give African farmers a break, there is ample reason to deal more fairly and cleanly with African governments when it comes to the continent's energy resources and minerals. World-market prices for commodities fluctuate, and declining earnings from metals that fueled exploitation by colonial powers have strained the economies of African states. Zambia, heavily reliant on copper, has seen its income shrink not only from reduced market prices but also because the landlocked country's most efficient outlet to the sea, via Angola, was destroyed by the prolonged conflict there. South Africa's economy has suffered from declining gold prices and increasing labor costs. But where one country ails, another thrives: Botswana's income from diamonds has soared, and a combination of stable, democratic government and judicious economic policy has moved Botswana into the middle-income category of African states, still a rare phenomenon.

The most significant development since decolonization, however, has been in energy resources and their exploitation. Africa is proving to contain far larger oil reserves than was estimated only 30 years ago, and today not only Nigeria (which supplies the United States with about 12 percent of its annual oil imports) and Angola but also Gabon, Chad, and southern Sudan are important producers.

The implications of Africa's rise among oil-producing regions is obvious in this time of strategic uncertainty and potential disruption in the Middle East. West and Equatorial African reserves lie directly across the Atlantic from the world's thirstiest consumer without choke-point perils or pipelines as vulnerable as those elsewhere. Thus the creation and maintenance of stable political and trade relationships between the United States and Africa's producers is in America's self-interest.

The implications for the citizens of the producing countries, however, are troubling. Peoples whose historic homes are in the oil-rich Niger Delta, for example, have waged a decades-long struggle to assert their rights to a far larger share of the proceeds than they have been awarded by rapacious regimes and voracious corporations. In 1995 the Abacha regime executed by hanging nine leaders of the Ogoni, a people in the vanguard of this movement, while a president-elect languished in jail. But neither trading-partner governments nor the oil corporations themselves risked their relations with Nigeria's dictatorial clique over this issue. To those Nigerians old enough to remember, it was all too familiar.

Events in the Niger Delta, in war-torn Angola (where oil and diamonds fueled the conflict), and in Sudan (where northerners ruthlessly dispossessed southern villagers beneath whose soil the oil was discovered) appear to have led to greater scrutiny of the actions of governments and oil companies. In 2000 the start of exploitation of Chad's oil reserve was accompanied by corporate compensation to rural citizens for their land and, according to EssoChad's reports, even for individual trees cut down to make way for power lines and part of the 665-mile pipeline to the coast of neighboring Cameroon (*Economist*, 2003). Further, an international agency will monitor the way the government of Chad spends the income it receives for its oil, the basis on which it awards contracts, and even the bank accounts of individual members. This effort to constrain corruption and protect citizens' rights may come to be regarded as a fresh start for oil-producing African states.

Whatever the outcome of this initiative, there can be no doubt that the world needs Africa's commodities. In a way, Africa is to America and Europe what Australia is to East and Southeast Asia: a storehouse of minerals not far, in global terms, from a vast market of consumers. But what Africa needs is what Australia has: a thriving economy of its own. In this respect the consumers of Africa's products must do more to help the source.

Africa matters to the rest of the world in still another way: as the global geographic realm most severely afflicted by dispersible diseases. Again the massive medical intervention Subsaharan Africa needs is not a matter of altruism: it is in the entire world's interest to improve public health in this, the sickest part of the planet.

What once was Africa's problem—the warming episode that made its tropics a vast incubator of a host of diseases—is now a global problem (Gould, 1993). Africa's links to the outside world may be weaker than those of other realms, but jet aircraft, passenger and cargo ships, and other conveyances can carry African maladies in human and animal vectors around the globe in a matter of hours (the first case of AIDS in North America may have been transmitted by an infected individual who traveled by jet from Kinshasa to Brussels to Chicago). The world is fortunate that a disease far more infectious than AIDS has not emanated from Africa to kill millions in short order, but medical geographers predict that such a time will come unless a huge, focused effort is made to improve public well-being here at the heart of the Land Hemisphere.

My colleagues who study such matters point out that dangerous diseases in the richer world for many years have received far more attention than diseases long confined to Africa. If malaria had been a midlatitude malady, they argue, it would have been wiped out long ago. From polio to smallpox, many rich-world diseases have been conquered. When AIDS took on serious dimensions in the United States, the medical establishment took it on full force, and today it is possible to keep HIV in check, perhaps indefinitely. Had AIDS remained confined to tropical Africa, it is unlikely that such a campaign would have been waged.

We know more about the endemic and epidemic maladies of Africa today because research is delineating them and because international agencies are reporting their prevalence, and because the media now vividly describe events such as outbreaks of the feared ebola fever and other dramatic but local health crises. African epidemics rarely become pandemics, as AIDS did, so the perceived threat of African infectious diseases spreading beyond Africa is low. But the risk is real and growing, and the best way to interdict it is to raise the level of well-being of Africans everywhere. Already, the World Health Organization (WHO) under UN auspices inoculates millions of African children against a range of diseases, but many more do not get the protection they need. As a result, infant and child mortality rates in many African countries rank among the highest in the world. Those who survive are likely to suffer from several of the endemic diseases of Africa against which protection is still inadequate.

Attacking Africa's diseases (much progress has been made, for example, against River Blindness) is one thing; making medicines affordable is another. As the AIDS disaster showed, even when Western drug companies had achieved remedies that can keep the virus at bay, the cost, while declining, remained too high for most Africans to afford. When the average annual income of a population is $600—the approximate figure for Tanzania—the

$300 monthly cost for the full range of needed drugs is simply out of the question. The subsidy proposed by President Bush in 2003 will go only part way to making AIDS medication available where it is needed most.

Improving the public-health situation in Subsaharan Africa should be a huge, international project not only to benefit Africans but also to reduce the risks Africa's medical situation poses to the rest of the world. In addition it is an indispensable element in any economic recovery Africa may make. The stakes are high and the potential returns immeasurable.

To the multicultural nation of the United States, Africa matters in still another way. The United States never held African colonies (its special relationship with Liberia arose from the repatriation of Africans who had come to America in bondage) but the African American population in this country is larger than the population of all but five Subsaharan African countries. In truth, Africa is America's second cultural source, after Europe, but the Americanization of the African minority has weakened the links between Africa and America's African diaspora. A recent press report suggested that more African Americans vacation in Europe than in Subsaharan Africa, and knowledge of Africa among African Americans appears to be no stronger than among whites.

This suggests a host of opportunities. In Brazil, which has a far stronger African imprint than the United States, interest in Africa is intense to the point that a major two-way flow of visitors has developed between the country of Benin (formerly Dahomey) in West Africa and the State of Bahia (centered on Salvador), following the discovery that many Africans were taken directly from Benin to Bahia during the period of slavery. The survival of African cultural traits in Bahia and the identification of ancestral homes in Benin created a remarkable Transatlantic cultural revival. For African Americans in the United States, such linkages may not be as discernable, but surely the recovery and well-being of Africa is a prospect of intense interest to all, and African American commitment and involvement would be key ingredients in the process. In this context, it is not difficult to hear the echo of former secretary of state Powell's inspiring words.

Africa's millennia of misfortune have not diminished the vibrancy of its cultures nor its importance in a world to which it gave cultural rise. For tens of thousands of years Africa nourished and forged humanity and launched emigrations that would forever change the planet. Africa's time and turn will come again.

EPILOGUE

The United States is the world's sole superpower at a momentous turn in global history. Concerns that were once remote and long range have become immediate and threatening, not just to America but to the world as a whole: rapid climate swings, great-power competition, terrorism, weapons of mass destruction in the possession of insurgents. All this is happening against a background of fundamental change in the growth and distribution patterns of the planet's now-huge human population, the prospect of an energy crunch that has yet to generate the necessary technological response, a rise of religious extremism not confined to Islam, and a persistent disparity between the prosperous and influential and the poor and powerless that, in this age of information, is clearer by the day even in the remotest parts of the world. Such notions as "international community" and "world system" can survive only by the acquiescence of the great majority, and from Palestine to Puntland to Papua this acquiescence is eroding.

Against such darkening clouds, where is the silver lining? When I watched the shocking events of 9/11 unfold and thought back to that terrible day in Rotterdam in 1940, I was reminded of the close calls this world has endured in the recent past and the very different course history might have taken. A cluster of racist European colonial powers might have achieved world domination a century ago were it not for Europe's historic divisiveness. If Hitler had waited for his weapons technologists to achieve in Germany what they did later in America and Russia and had launched his war with a monopoly on jet aircraft and long-range missiles, neither the English Channel nor the North Atlantic would have saved the world. Had the inefficiencies, contradictions, and self-destructiveness of Moscow's version of communism not constrained the Soviet Union, Moscow might be at the center of a global rather than a regional empire today, and the gulag would extend around the world. Given that great-power competition is an apparently inevitable culmination of the human drive for dominance, it can be argued that the world was fortunate that, among such contestants, it was the United States that emerged as the leading contender. In the immediate aftermath of the Second World War the United States, with its WMD capacity, was in a position to impose its will upon the world. It did not do so, but it is difficult to imagine another power with similar advantage acting with similar restraint. Instead, the United States enabled European countries to reconstruct their economies even as it promoted the deconstruction of colonial empires, it brought federal democracy to Germany even as its forces occupied that defeated enemy, and it introduced representative government while guiding economic recovery in Japan only four years after the Pearl Harbor attack often cited in the context of 9/11. One of America's proudest traditions lies in its ability to ignore the painful past, eschew revenge, and build for the future.

Which is not to suggest that the United States had not, over the past half century, contradicted its traditions and principles. While espousing decolonization and democracy it had common (Cold War) cause with some of the world's worst tyrants. In the name of containing communism, a sound cause indeed, the United States intervened, overtly and covertly, in countries that appeared poised to experiment with it. While urging adversaries to respect the rights of their minorities, Washington failed to intervene when "friendly" regimes engaged in genocide. And now, United States forces are trying to contain "collateral damage" in a country that posed no immediate threat to America—collateral damage that, according to a study published in the journal *Lancet* in November 2004, included 100,000 Iraqi civilian deaths even before the end of that year. Add to this the events in United States–controlled prisons in Iraq and the legal issues arising from the incarceration of terrorist suspects at Guantanamo, and there is still other collateral damage to be concerned about.

Despite two years of warfare and insurgency, Americans still lack adequate insight into Iraq's diverse cultural fabric and regional mosaic, and their mental maps of Iraq continue to be distorted. The longstanding belief that "war teaches geography" simply doesn't appear to apply. If North Vietnam and South Vietnam were hazy concepts four decades ago, then Iraq's regional-cultural layout (let alone its neighbors') are vaguer still in the public eye. That means that the risks of failure are misunderstood and the consequences of success (in whatever form that will de declared) are overestimated. When United States leaders cite the Marshall Plan and Japan's revival as campaigns comparable to Iraq, the only remaining question is whether such assertions constitute deliberate distortion or genuine geographic ignorance. Neither is an encouraging option.

The United States faces crucial challenges in the world today, and bringing democracy to one Muslim country with a tradition of authoritarianism and violent retribution is not one of them. Iraq will be a diversion unless the insurgency American intervention generated spreads and overtakes terrorism in the Middle East beyond Iraq's borders, which remains a possibility. Under such circumstances, the experiences in the Sunni Triangle will be repeated on a far larger scale. A domino effect in Saudi Arabia or other Iraqi neighbors and the destabilization of other regimes in the region, still an al-Qaeda objective, would conjoin the terrorist and insurgency threats into something much more difficult to contain.

American (and other countries') dependence on oil and natural gas drawn from reserves in Islamic countries threatened by instability thus creates a vulnerability with geopolitical implications. In late 2004, when the cost of a barrel of oil exceeded $55 and seemed headed for $60, the potential impact of a major interruption of supplies from Saudi Arabia and elsewhere was reflected by the financial markets, and it was only a shortage, not any real interruption, that caused those fears. The largest reserves of oil lie in Saudi Arabia, Iraq, the United Arab Emirates, and Kuwait (the largest gas reserves lie in Russia), and Iraq's production has already been diminished by a series of acts of sabotage on installations and pipelines. Al-Qaeda's plotters know that a successful strike on Saudi Arabia's facilities would deal a crucial blow not only on Saudi income but also on Western economies.

There are other energy worries that should cause the United States to invoke all its technological capacities in a search for alternative energy sources. Large as the known reserves are, they are declining, and China's fast-growing demand will speed that process up. Smaller reserves outside the Muslim world are being depleted rapidly as well: North Sea and Gulf of Mexico yields are dropping fast. Undoubtedly new reserves will be found, but they will lie in ever-more remote locales on land (Siberia perhaps) and at sea (below deeper waters off Africa and South America). This increases the cost of discovery and exploitation, and some day $60/barrel oil may be a long-lost bargain.

For America's national security, we need an energy revolution, a Manhattan II Project (what a felicitous name, under the circumstances) that will do for energy what the original Manhattan Project more than 40 years ago did for armaments. Whether our future fuel will be based on hydrogen or derived from nuclear fission or fusion, the Western world must find a way to reduce and minimize its dependence on fossil fuels. We have allowed ourselves to be restrained from converting to nuclear power plants (largely on the grounds of risk, but tell that to the families of the hundreds of miners in Asia and Africa and elsewhere who die each year in mines producing coal for those "safe" coal-burning plants). We need alternative energy technology, and we are able to achieve it—but the cost will be high. To accomplish it, we need leadership that is willing to look the country in the eye and say that we have no choice, leaders who will tell the people that security comes at a price, leaders who will put the nation before oil companies when it comes to difficult decisions. Americans still pay about one-third of what Europeans pay for a gallon of gas. European highways are not exactly deserted. America can absorb a graduated, fractional rise in the cost of a gallon of gas, dedicated to the Manhattan II Project, and politicians should take a vow not to accuse each other of "raising taxes."

The original Manhattan Project produced its key result—sustained nuclear reaction—almost exactly one year after Pearl Harbor. It has been nearly four years since September 11, 2001. Well, oil dividends are huge.

It is a great deal more difficult to prepare for abrupt climate change than it is for energy shortages. After all, we have had some forewarnings where energy is concerned. Some of us remember those gas lines during the Carter administration, when roads were clogged with double lines of cars that went around the block and beyond, fistfights started when cars broke into line, gas stations limited customers to 10 gallons of gas or fewer, pumps sometimes ran dry for people who had sat in line for three hours or more. When it transpired that the United States sold grain to Muslim countries that limited their exports, bumper stickers read "A Bushel for a Barrel" and public opinion wanted the grain exports stopped.

The last climate warning came centuries ago, and even if we could remember the details would not be relevant: the Earth's population was just a fraction of what it is today. But locally, the impact was devastating. In Western Europe, fertile farmlands lay waste, destitute farm families crowded into the cities where disease was rampant and death rates skyrocketed. Storage facilities of the modern kind did not exist; markets lay bare and hunger stalked the land. Searing, bone-dry summer heat followed bitter winter cold. Huge storms drove the seas over dikes and into polders.

The Dutch learned to deal with such contingencies, building ever-more sophisticated coastal protection systems in the form of giant steel doors against rising waters; the British constructed an elaborate project across the Thames to stop a surge from reaching London. But in truth, as the 2004 hurricane season reminded Floridians, it is impossible to prepare adequately for the wrath of nature. Damage will be done, and coastal areas will suffer most.

Yet something *can* be done to mitigate the most serious effects of a climate catastrophe of the kind Western Europe experienced around the turn of the fourteenth century. People would be driven from their homes and away from coastal areas, and there would be massive dislocation, but such is the housing stock of the nation today that most if not all could be sheltered. The real crisis would be over food and drinkable water. And in this arena, action must be taken.

During the Cold War with the Soviet Union, the prospect of a global nuclear conflagration was real and many homeowners added bomb shelters to their properties, some elaborately lined with antiradiation insulation and stocked with several months' worth of food. Perhaps more importantly, the government began a massive program of grain and canned food storage, and some of my graduate students were involved in the location of the storage facilities and the design of the distribution system should it be needed. The plan would undoubtedly have been expanded over time, but the decline and fall of the Soviet Union rendered it obsolete. Now is the time to revive and reconstruct it. If what we know about the start of the Little Ice Age is accurate, abrupt environmental change is preceded by a period of wild climatic fluctuations. These fluctuations have maximum impact in higher latitudes and along coastlines, which is where the preparations are most urgent. Over the short term, new ways and routes to provide what the country needs will be found, but any immediate crisis must be alleviated through food storage and other preparations.

Make no mistake: abrupt climate change constitutes a threat to national security. It has the capacity to disrupt order throughout the world and to level the strategic playing field. Studies of this aspect of the issue urge the creation of a crisis-response mechanism that would coordinate the responsibilities of all military and security operations, from the armed forces to the coast guard, and from Homeland Security to FEMA, for activation in the event of an environmental emergency. Imagine what the impact would be of a Tambora event on the world today, let alone Toba, still minor incidents compared to abrupt climate change. The tsunami disaster on December 26, 2004, and the ensuing difficulties in mounting emergency relief and medical responses illustrate the importance of preparedness. We must have plans in place as a matter of national security.

What can be done to avert a potential confrontation with China? This challenge will evolve incrementally, but the first order of business clearly involves Taiwan. The United States must press Taiwan's leaders not to pursue their notions of independence to the point of precipitating a crisis, while at the same time warning Beijing of the incalculable consequences of any attempt to use force to conquer the island. China's installation of batteries of missiles aimed at Taiwan is troubling, since the Chinese are well aware of United States commitments to protect the Taiwanese. Time will soften the rough edges of this problem, and already has: two decades ago it was inconceivable that Taiwanese investment in China's Pacific Rim economy would even be possible, let alone reach the dimensions it has. An outbreak of armed conflict would be catastrophic for all three parties. Nothing that increases this risk should be tolerated, and Washington must convince Beijing that this must be a joint objective.

Over the longer term, China's economic and geopolitical challenge will be more difficult to accommodate, and the American role in East Asia will undoubtedly have to change. China has unresolved issues with Japan ranging from Japan's failure to acknowledge the atrocities it committed in China during its wartime occupation to disputes over islands and waters in the South China Sea. These issues have remained relatively subdued because of the constitutional restrictions placed on Japan's military and because the United States has kept armed forces based in Japan. But there are signs that Japan will loosen the constraints on its military power, largely because of security concerns arising from North Korea's nuclear ambitions and maritime incursions, and the American government is contemplating a reduction if not elimination of its military presence in Japan. Those developments will change the geopolitical map of East Asia. As was suggested earlier, other circumstances are also changing the relationships between China and the United States, and let there be no doubt: the greater America's public awareness and knowledge of China and its aspirations, the better the chances that understanding will mitigate confrontation.

In its role as global superpower, the United States faces numerous other problems from which it cannot turn away: helping the Palestinians and Israelis reach a territorial accommodation (a geographic problem if there ever was one); joining the Europeans in seeking a solution to Iran's nuclear ambitions; leading a multinational effort to inhibit North Korea's nuclear-weapons development; repairing frayed relations with the European Union and especially its key mainland members; taking a constructive role in the reorganization of the United Nations that is badly needed.

Most of all, the United States must lead by example, by adhering scrupulously to its own principles and traditions and by taking actions to conform to its words. Compromise is unavoidable in a complex world, and it would be impractical to terminate all contact with dictatorial regimes. But it would indeed be possible to eliminate all remaining tariffs, subsidies, and quotas and thus to live up to the free-trade ideals Washington so vocally embraces. It would also behoove the United States to recognize more demonstratively how fortunate it is to have Canada and Mexico as neighbors as well as NAFTA partners. Canada may have its disagreements with the United States on some issues, but relations between Washington and Ottawa are generally good even when Canadian public opinion suggests otherwise. In an era when lost territory arouses passions ranging from irredentism to terrorism, Mexico is a friendly and cooperative neighbor whose president, Vicente Fox, exhibited remarkable statesmanship when the much-anticipated "special relationship" with Spanish-speaking President Bush turned into something rather less than that. And Americans who complain about Mexican immigration might consider the alternative faced by Spain and Western Europe.

To prolong the "American century" and to bestow on the world the benefits of American leadership, the United States must rebuild its own foundations. This is the world's second-largest democracy, but its democratic machinery, from electoral equipment to the Electoral College, is creaking. The United States, given its size, is the world's most successful multicultural society, but given its wealth it could do much better than income disparities indicate. America's Constitution guarantees freedom of (and from) religion, but fundamentalist extremism is rising. Freedom of speech and the unpunished expression of unpopular opinion is a hallmark of America, but there are disturbing new signs of intimidation. The United States is also living beyond its means, a troubling negation of the principles of good management that have made this country, for most of its existence, strong as well as confident.

As America's allies (not to mention its adversaries) never tire of saying, the United States in recent years has alienated much of the world through unilateral and overbearing actions in pursuit of goals with which many of its cohorts do not agree. But in countless contexts, from keeping the peace in East Asia to making peace possible in Eastern Europe, from combating AIDS in Subsaharan Africa to pressing Sudan on the rights of its Southerners, from extending free trade in the Americas to ending the nuclear threat in rogue-state Libya, the United States has played a key role in making this a better world. Certainly the United States could and should do better; in terms of per capita aid to poorer countries, for example, the American government ranks near the lowest among the rich countries (but private donations are among the most generous). The United States' decision not to participate in the Kyoto, UNCLOS, International Criminal Court, and other current multinational initiatives may be justifiable in the context of their particular impact on this superpower, but to the world it appears to place national interests before global concerns. Inconsistencies in the pursuit of doctrines regarding freedom and democracy create worldwide political uncertainties on which America's adversaries are capitalizing.

Superpowers tend to misbehave, from imperial subjugation to ideological compulsion to strategic blackmail. On balance, it is nevertheless difficult to contemplate a world without the American superpower presence.

The geographies of Germany and Korea with their stark west-east, south-north contrasts provide the objective regional evidence—if not of good and evil, then at least of better and worse. In the aftermath of World War II, the United States was the indispensable global presence. The transformation of China and the reconstitution of Russia, unfinished though they are, were enabled by America's stabilizing influence. The capacity of the United States to meet the challenges now looming will shape the future of this dangerous world.

WORKS CITED

9/11 Commission, 2004. *Final report of the National Commission on Terrorist Attacks Upon the United States.* New York: Norton.

Alley, R. B., 2002. *Abrupt Climate Change.* Washington, D.C.: National Academy Press.

Altman, L. K., 2002. "By 2010, AIDS May Leave 20 Million Orphans." *New York Times,* July 10.

Annan, K., 2004. "Courage to Fulfill Our Responsibilities." *Economist,* 12/4: 21.

Ardrey, R., 1966. *The Territorial Imperative.* New York: Atheneum Press.

Aryeetey-Attoh, S., Ed., 1997. *Geography of Subsaharan Africa.* Upper Saddle River: Prentice-Hall.

Baker, J. H., 1996. *Russia and the Post-Soviet Scene: A Geographical Perspective.* Hoboken: John Wiley & Sons.

Begun, D. R., 2003. "Planet of the Apes." *Scientific American,* 289: 2.

Best, A. C. G., and H. J. de Blij, 1977. *African Survey.* New York: Wiley.

Charlemagne, 2003. "Europe's Population Implosion." *Economist,* 7/19: 42.

Clarke, R. A., 2004. *Against All Enemies.* New York: Free Press, 40.

Cohen, J. E., 2003. "Human Population: The Next Half Century." *Science,* 302: 14, 1172.

Curtin, P., 1969. *The Atlantic Slave Trade.* Madison: University of Wisconsin Press.

Cutter, S. L., D. B. Richardson, T. J. Wilbanks, Eds., 2003. *The Geographic Dimensions of Terrorism.* New York: Routledge.

Davis, C. S., 2004. *Middle East for Dummies.* Hoboken: John Wiley & Sons.

Davis, S., 2002. *The Russian Far East: The Last Frontier.* New York: Routledge.

de Blij, H. J., 1971. *Geography: Regions and Concepts.* New York: Wiley.

de Blij, H. J., 1991. "Africa's Geomosaic Under Stress." *Journal of Geography,* 90: 1.

de Blij, H. J., 1995. *Harm de Blij's Geography Book.* New York: John Wiley & Sons.

de Blij, H. J., 1996. "Tracking the Maps of Aggression." *New York Times,* October 28.

de Blij, H. J., 2003. "Seeking Common Ground on Iraq." *New York Times,* October 11.

de Blij, H. J., and P. O. Muller, 2004. *Geography: Realms, Regions and Concepts,* 11th ed. Hoboken: John Wiley & Sons.

de Blij, H. J., 2004. "Africa's Unequaled Geographic Misfortunes." *Pennsylvania Geographer* 42/1: 3–28.

de Blij, H. J., Ed., 2005. *Atlas of North America.* New York: Oxford University Press.

Demko, G. J. et al., Eds., 1998. *Population Under Duress: The Geodemography of Post-Soviet Russia.* Boulder: Westview Press.

Diamond, J., 1997. *Guns, Germs, and Steel.* New York: Norton.

Diamond, J., 2005. *Collapse: How Societies Choose to Fail or Succeed.* New York: Viking.

Economist, 2003, "Can Oil Ever Help the Poor?" 12/4: 42.

Economist, 2004a, "The Ultra-Liberal Socialist Constitution," 9/18: 59.

Economist, 2004b, "Why Europe Must Say Yes to Turkey," 9/18: 14.

Economist, 2004c, "The Impossibility of Saying No," 9/18: 30.

Fagan, B., 2000. *The Little Ice Age*. New York: Basic Books.

Friedman, T., 1996. "Your Mission, Should You Accept It." *New York Times*, October 27.

Gewin, V., 2004. "Mapping Opportunities." *Nature*, 247: 376.

Gould, P. R., 1993. *The Slow Plague: A Geography of the AIDS Pandemic*. Cambridge: Blackwell.

Graham, R., and J. Nussbaum, 2004. *Intelligence Matters*. New York: Random House.

Grove, J. M., 2004. *The Little Ice Age*, 2nd ed. London: Methuen.

Hall, S. S., 1993. *Mapping the Next Millennium*. New York: Random House.

Hansen, J., 2004. "Defusing the Global Warming Time Bomb." *Scientific American*, March, 68–77.

Harris, S., 2004. *The End of Faith: Religion, Terror, and the Future of Reason*. New York: Norton.

Harvey, D., 2004. *The New Imperialism*. Oxford: Oxford University Press.

Harvey, R., 2003. *Global Disorder: America and the Threat of World Conflict*. New York: Carroll & Graf.

Hertslet, E., 1909. *The Map of Africa by Treaty*. London: His Majesty's Stationery Office.

Hitchcock, W. I., 2002. *The Struggle for Europe*. New York: Doubleday.

Horgan, J., 1996. *The End of Science: Facing the Limits of Knowledge in the Twilight of the Scientific Age*. Reading: Addison-Wesley.

Huntington, E., 1942. *Principles of Human Geography*. New York: Wiley.

Huntington, E., 1945. *Mainsprings of Civilization*. New Haven: Yale University Press.

Huntington, S. P., 1996. *The Clash of Civilizations and the Remaking of the World Order*. New York: Simon & Schuster.

Kagan, R., 2004. *Of Paradise and Power*. New York: Vintage.

Kaiser, R. J., 1994. *The Geography of Nationalism in Russian and the U.S.S.R.* Princeton: Princeton University Press.

Kaplan, R., 2000. *The Coming Anarchy*. New York: Random House.

Kasting, J. F., 2004. "When Methane Made Climate." *Scientific American*, 291/1: 78.

Kepel, G., 2002. *Jihad: The Trial of Political Islam*. Cambridge: Harvard University Press/Belknap.

Kepel, G., 2004. *The War for Muslim Minds and the West*. Cambridge: Harvard University Press/Belknap.

Kerr, R. A., 2003. "Has an Impact Done it Again?" *Science*, 302: 21.

Langewiesche, W., 2004. *The Outlaw Sea*. New York: North Point Press.

Lincoln, B., 1994. *The Conquest of a Continent: Siberia and the Russians*. New York: Random House.

Mackinder, H. J., 1904. "The Geographical Pivot of History." *Geographical Journal*, 23: 421–437.

Martin, G., 2005. *All Possible Worlds: A History of Geographical Ideas*. New York: Oxford University Press.

Monmonier, M., 1991. *How to Lie With Maps*. Chicago: University of Chicago Press.

Monmonier, M., 1997. *Cartographies of Danger*. Chicago: University of Chicago Press.

Muehrcke, P. C., and J. O. Muehrcke, 1997. *Map Use: Reading-Analysis-Interpretation*, 4th ed. Madison: JP Publishers.

Murck, B. W. and B. J. Skinner, 1999. *Geology Today*. New York: John Wiley & Sons.

Murdock, G. P., 1959. *Africa: Its Peoples and Their Culture History.* New York: McGraw-Hill.

Myers, S. L., 2004. "Putin Gambles on Raw Power." *New York Times,* September 19.

National Geographic, 2005. *Atlas of the World,* 8th ed. Washington, D.C.: National Geographic Society.

Newman, J., 1995. *The Peopling of Africa: A Geographical Interpretation.* New Haven: Yale University Press.

Onishi, N., 2001. "Rising Muslim Power in Africa Causes Unrest in Nigeria and Elsewhere." *New York Times,* November 1.

Oppenheimer, S., 2003. *The Real Eve: Modern Man's Journey Out of Africa.* New York: Carroll and Graf.

Oxford University Press, 2004. *Atlas of the World,* 12th ed. New York: Oxford University Press.

Patten, C., 1998. *East and West: The Last Governor of Hong Kong on Power, Freedom, and the Future.* London: Macmillan.

Qiang, S., 1999. *China Can Say No.* Beijing: National Press.

Rachman, G., 2004. "Outgrowing the Union." *Economist,* Survey of the European Union, 9/25: 3.

Reader, J., 1998. *Africa: Biography of a Continent.* New York: Knopf.

Remnick, D., 1993. *Lincoln's Tomb: The Last Days of the Soviet Empire.* New York: Random House.

Rifkin, J., 2004. *The European Dream.* New York: Tarcher/Penguin.

Robinson, A. H., and B. Petchenik, 1976. *The Nature of Maps: Essays Toward Understanding Maps and Mapping.* Chicago: University of Chicago Press.

Rother, L., 2002a. "Argentine Judge Indicts Four Iranian Officials in 1994 Bombing of Jewish Center." *New York Times,* March 10.

Rother, L., 2002b. "South America Under Watch for Signs of Terrorists." *New York Times,* December 15.

Sachs, J. D., 2000. *The Geography of Economic Development.* Newport, RI: United States Naval War College Jerome E. Levy Occasional Paper in Economic Geography and World Order.

Schmitt, E., and T. Shanker, 2004. "Estimates by U.S. See More Rebels with More Funds." *New York Times,* October 22.

Shaw, D. J. B., Ed., 1999. *Russia in the Modern World: A New Geography.* Malden: Blackwell.

Solis, P., 2004. "AAG Member Profile: E. 'Fritz' Nelson." *AAG Newsletter,* 39/1: 9.

Spykman, N. J., 1944. *The Geography of the Peace.* New York: Harcourt, Brace & Co.

Stanley, W. R., 2001."Russia's Kaliningrad: Report on the Transformation of a Former German Landscape." *Pennsylvania Geographer,* 39: 1.

Stock, R., 2004. *Africa South of the Sahara: A Geographical Interpretation,* 2nd ed., New York: Guilford Press.

Terrill, R., 2003. *The New Chinese Empire: What it Means for the United States.* New York: Basic Books.

Thornton, J., and C. E. Ziegler, Eds., 2002. *Russia's Far East: A Region at Risk.* Seattle: University of Washington Press.

Tishkov, V. 2004. *Chechnya: Life in a War Torn Society.* Berkeley: University of California Press.

Trenin, D., 2002. *The End of Eurasia: Russia on the Border Between Geopolitics and Globalization.* Washington: Carnegie Endowment for International Peace.

Van Natta Jr., D., and L. Bergman, 2005. "Militant Imams Under Scrutiny Across Europe." *New York Times*, October 25.

Wegener, A., 1915. *The Origins of Continents and Oceans*. New York: Dover (reprint of the 1915 original, trans. by John Biram, 1966).

Wilford., J. N., 1981. *The Mapmakers*. New York: Knopf.

Wilson, E. O., 1995. *Naturalist*. Washington, D.C.: Island Press.

Wright, R., 1986. "Islamic Jihad." Encyclopaedia Britannica Book of the Year.

INDEX

Pakistan (*continued*)
  population, *99*
  secession, 112
Paleocene era, *59, 63–64, 64*
Paleozoic era, 60, 61
Palestine
  Christianity in, 164
  immigration, 155
  and Iraq, 189
  and Israel, 50, 151, 152, 161, 163, 175, 280
Palestinian Liberation Organization (PLO), 153
Panama, 146, *180*
Pangaea, 54–55, *59*, 60, *61*
Papua, 3
Paraguay, 176, *180*
parallels, 31
Paris, France, 177
Partnership and Cooperation Agreement, 253
Patten, Christopher, 138
Patterns of Global Terrorism (U.S. State Department), 161
Peace Corps, 15, 266
Pennsylvanian period, *59*
People's Republic of China, 136. *See also* China
Permian Ice Age (or Dwyka Ice Age), *59*, 60, 61, 62
Perot, Ross, 209
Persian Gulf (Arabian Gulf), 38–39, 70, 128, *162*
Peru, 120, *180*
Peshawar, Pakistan, 176
Petchenik, Barbara B., 24
Peter I, 236
petroleum. *See* oil and oil reserves
Philippines, 106, 129, 145–46, 178
Phuket, Thailand, 12, 178
Physical Geography of China (Sonqiao), 44–45
Pinyin system, 37
Pirtle, Charles, 16
Pitman, Walter, 75
Pivot Area (Heartland Theory), 128
Planet Hollywood bombing (1998), 187
plate tectonics, 54–57, 64
Pleistocene era, *59, 64*, 65–72, 78, 256, 259
Pleistocene interglacials, 73, 75, 82
Pliocene era, *59, 64*, 65, 67, 68
Poland
  borders and boundaries, 120–21
  devolutionary pressure, *206*
  and European Union, 212, 213, *218*, 224–26
  and Russia, 231, 253
polar ice, 52. *See also* glaciers
political science, 11
politics, 13, 14
pollution, 47, 52, 115, 134
Pol Pot, 112
Polynesians, 102
population, **91–107**. *See also specific regions and countries*

distribution, *98–99*
"doubling times" for, 105
and environment, 101–3
future patterns, 95–97
and global warming, 53, 82, 278
growth, 51, 92–95, *94*
on maps, 116–17
overpopulation, 92, 100, 101
and politics, 100–101
ZPG (Zero Population Growth), 93, 106
Portugal
  and China, 136, 142
  colonialism, 262, 263, 264
  and European Free Trade Association, 210
  and Muslim realm, 161
  nation-state model, 110
Postglacial Optimum, 76
Potsdam Agreement, 224
poverty, 8–9, 256
Powell, Colin, 256, 274
Precambrian, *59*
primates, 65–68, *68*, 72
Primorskiy Territory, Russia, 242, 243
Proterozoic eon, 58, *59*, 60
Protestant Reformation, 166
provincialism, 21
public transportation, 203
Pudong, China, 129, 139
Puntland, 37–38, 185
Pushtuns, 157–58
Putin, Vladimir, 247, 250–51, 252, 256

Qing (Manchu) Dynasty, *135*, 135–36, 138
Quebec, Canada, 109
Queen Elizabeth Islands, 108
Quran, 164

RAND Corporation, 175
Ratzel, Friedrich, 111
Reader, John, 264
Reagan, Ronald, 17, 50
Real Irish Republican Army (RIRA), 161
Red Guards, 126
Red Sea, 70, 259
refrigeration, 93
reglaciation, 78
Reilly, Jack, 18
relative location in maps, 33
religion. *See also specific religions, including* Islam
  and European Constitution, 214–15
  extremism, 21
  freedom of religion, 166, 281
  and population, 94
  religious conversion, 165
  separation of church and state, 166
remote sensing, 46
reptiles, 60, 62
Republic of Ichkeria (Chechnya), 247
Republic of Somaliland, 37